KING

Published by the Press Syndicate of the University of Cambridge
The Pitt Building, Trumpington Street, Cambridge CB2 1RP
32 East 57th Street, New York, NY 10022, USA
296 Beaconsfield Parade, Middle Park, Melbourne 3206, Australia

First published 1983

Printed in Great Britain by Belmont Press, Northampton

Library of Congress catalogue card number: 81-21784

British Library Cataloguing in Publication Data

Redfern, Margaret
Insects and thistles. – (Naturalists' handbooks; 4)

1. Insects 2. Thistle
I. Title II. Series
595.7 QL463

ISBN 0 521 23358 5 hard covers
ISBN 0 521 29933 0 paperback

Naturalists' Handbooks 4

Insects and thistles

MARGARET REDFERN

With illustrations by
ANTHONY J. HOPKINS

Cambridge University Press
Cambridge
London New York New Rochelle
Melbourne Sydney

Editors' preface

Sixth formers and others without a university training in biology may have the opportunity and inclination to study local natural history but lack the knowledge to do so in a confident and productive way. The books in this series offer them the information and ideas needed to plan an investigation, and the practical guidance needed to carry it out. They draw attention to regions on the frontiers of current knowledge where amateur studies have much to offer. We hope the readers will derive as much satisfaction from their biological explorations as we have done.

The keys are an important feature of the books. Even in Britain, the identification of many groups remains a barrier to ecological research because experts usually write keys for other experts, and not for general ecologists. The keys in these books are meant to be easy to use. Their usefulness depends very much on the illustrations, the preparation of which was assisted by a grant from the Natural Environment Research Council.

S.A.C.
R.H.L.D.

Acknowledgements

I should like to thank several people who have lent specimens and helped with advice, especially Mr David Carter (moths), Dr Victor Eastop (aphids), Dr Dennis Unwin (flies) and Dr Mike Morris (weevils), and my husband, Dr Robert Cameron, who commented helpfully on the manuscript.

M.R.

1 Introduction

Thistles are common plants found throughout the British Isles, often invading crops and meadows as weeds. Two species, *Cirsium arvense* Creeping thistle and *C. vulgare* Spear thistle, are discussed here and the word 'thistle' applies particularly to these species.

Despite their spines, thistles support a rich and varied fauna of insect herbivores with their associated predators, parasites and inquilines. Thistles have spread rapidly in countries to which they have been introduced without their insect fauna, often reaching pest status, suggesting that the fauna plays an important part in limiting their numbers (P. Harris, 1973;* and see chapter 5).

inquiline: an animal living in the home of another species, and using its food

Most species dependent on thistles are herbivores. Most are restricted to particular microhabitats on the thistle and there are distinct guilds of insects associated with different parts of the plant. The flower heads, the insides of stems, roots and leaves carry characteristic concealed faunas of gall-inducers, borers and miners, apart from the more conspicuous insects which visit flowers, browse leaves or suck sap. Within each guild there are differences in timing of life cycle, methods of feeding and causes of mortality which may minimise competition. Some species are monophagous and others are highly polyphagous, but most show varying degrees of specialisation between the two extremes, being found on a number of species of Compositae (Zwölfer, 1965).

monophagous: feeding on only one species of plant

oligophagous: feeding on only a few species of plant

polyphagous: feeding on many species of plant

Carnivores are less restricted, feeding on a range of insect species and selecting microhabitats which may occur on many plant species. Insect parasites (parasitoids) are more conservative; each develops on a single host insect, and most species are restricted to a few related host species, particularly those with similar habits.

parasitoid: an insect parasite that feeds as a larva on a single host insect

The plants themselves provide a succession of microenvironments. Communities associated with the flower heads develop during summer, detrital feeders colonise the brown, dead heads during late autumn and winter, and more enter when the heads and stems have fallen.

Thistles contain a number of independent communities, connected only by their common dependence on the plant and by the relatively few roving predators which wander all over the plant. These miniature communities offer opportunities for studying a variety of ecological and entomological problems, and the fauna of

*References cited under authors' names in the text appear in full in Further Reading on p. 63.

parts of thistle plants (or closely related species such as knapweeds) has been used to study food webs and life cycle interactions (Redfern, 1968), energetics (Cameron & Redfern, 1974), population dynamics (Varley, 1947), competition and parasite–host interactions (Zwölfer, 1970, 1979).

Because of the pest status of thistles in many countries, many of their herbivorous insects have been examined for use as agents of biological control (Zwölfer & Harris, 1966; Zwölfer, 1968, 1969; Claridge, Blackman & Baker, 1970; P. Harris, 1973; Peschken & Beecher, 1973), and these studies have increased our knowledge of their ecology. There is a continuing need for information on insects of potential value in biological control, and studies of species common on thistles in Britain can therefore be valuable.

biennial: life cycle of 2 years

perennial: life cycle of many years

To identify *Cirsium vulgare* and *C. arvense* (pl. 1), see Clapham, Tutin & Warburg (1958) and Fitter, Fitter & Blamey (1974). Relatives of these thistles are shown in table 1. *C. vulgare* is biennial and reproduces by seed. *C. arvense* is perennial and most reproduction is vegetative, although seed is viable and, blown by wind, can colonise new areas. Buds of *C. vulgare* (fig. 1) are spiny, unlike the smaller *C. arvense* buds (fig. 2).

Fig. 1. *C. vulgare* bud.

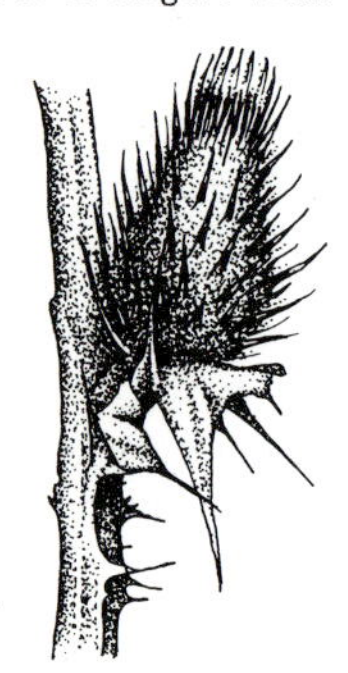

Somewhere in any scientific publication, each scientific name should appear in full, together with the name of the authority, the person who originally named the species. In this book authorities are given in the keys in full except for L. (Linnaeus), F. (Fabricius), D. & S. (Dennis and Schiffermueller) and R.-D. (Robineau-Desvoidy). Names change as a result of taxonomic research, and it may be necessary to refer to a checklist, such as Kloet & Hincks (1964–78) for insects, to find the modern equivalent of an out-of-date name.

In the following chapters all botanical names follow Clapham *et al.* (1958) and insect names Kloet & Hincks (1964–78, except for the weevil *Cleonis* (= *Cleonus*) *piger*.

Fig. 2. *C. arvense* bud.

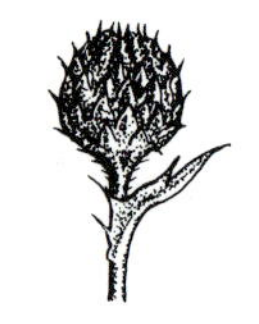

Table 1. *Subtribes of thistles and their relatives in the tribe Cynareae, family Compositae*

CARDUINAE	CENTAUREINAE
Arctium spp. (burdocks)	*Saussurea alpina* (L.) DC.
Carduus spp. (thistles)	*Centaurea* spp. (knapweeds)
e.g. *C. nutans* L. (Musk thistle)	*Serratula tinctoria* L. (Saw-wort)
Cirsium spp. (thistles)	*Carthamus* spp.[a]
e.g. *C. vulgare* (Spear thistle)	
C. arvense (Creeping thistle)	CARLINAE
C. palustre (Marsh thistle)	*Carlina vulgaris* L. (Carline thistle)
Silybum marianum (L.) Gaertn. (Milk-thistle)	
Onopordum acanthium L. (Cotton thistle)	ECHINOPINAE
	Echinops spp.[a] (Globe thistles)

[a]*Carthamus* and *Echinops* are not native in Britain.
After Clapham *et al.* (1958), Zwölfer (1965).

2 The resident fauna of thistle heads

achene: fruit containing single seed at base of floret

frass: caterpillar faeces

detritivore: an animal that feeds on detritus

Flower heads of thistles contain the most varied, specific and well-known insect fauna of any part of the plant. They are a rich source of food, packed with achenes, and their inhabitants are protected from vertebrate predators by the tough spiny bracts, especially in *C. vulgare*. The larvae of several herbivorous groups live here – tephritid flies, cecidomyiid midges and moths – together with their parasites. After death, the flower heads and miscellaneous frass and animal debris contained within them offer habitats to a diverse detritivore community.

Tephritidae (picture-winged flies)

Many tephritids are recorded (table 2), some of which are occasional strays usually found on other Cynareae, especially *Centaurea* (knapweeds) and *Carduus* (other thistles). Some feed as larvae on achenes and receptacle tissue, protected by a cocoon of floret hairs; others induce galls. *Urophora stylata* is the commonest galling species and the one which has the greatest effect both on the thistle and on the associated fauna. It is usually restricted to *C. vulgare* and, because it is the best known herbivore, it is discussed in some detail. *Terellia serratulae* is the commonest non-galling tephritid.

Table 2. *Tephritid flies occurring as larvae in flower heads of* Cirsium vulgare *and* C. arvense *(pl. 1)*

Species	Specificity	Main host plant	Gall formed
Chaetostomella onotrophes	*	*Centaurea nigra*	No
Chaetorellia jaceae	*	*C. nigra*	Yes
Terellia serratulae	**	*Cirsium, Carduus*	No
Orellia ruficauda	**	*Cirsium arvense* *C. palustre*	No
O. winthemi	**	*Carduus crispus*	No
Xyphosia miliaria	**	*Cirsium arvense*	No
Urophora solstitialis	*	*Carduus crispus* *C. nutans*	Yes
U. stylata	**	*Cirsium vulgare*	Yes
Tephritis cometa	***	*Cirsium arvense*	No
T. conura	**	*Cirsium*	Yes
Acanthiophilus helianthi	*	*Cirsium, Centaurea*	Yes

***Monophagous, specific to a single plant species.
** Oligophagous, found only on plants in same subtribe Carduinae (table 1).
* Polyphagous, found on Carduinae and other subtribes of Cynareae (table 1); sometimes on other Compositae.

Information on distribution is scanty; most records are from S. England where most collecting has been done; an exception is *Tephritis conura*, commoner in N. England and Scotland.

From Séguy (1934), Niblett (1939, 1953, 1956), Varley (1937), Persson (1963), Buhr (1964), Zwölfer (1965), Colyer & Hammond (1968), Redfern (1968), Moore (1975), Stubbs & Chandler (1978), Unwin (1981).

Urophora stylata

Life cycle

The life cycle of *Urophora stylata* is characteristic of gall-inducing tephritids in thistles (fig. 3). Adult flies (pls. 2.1, 2.2) emerge in June and July and fly until mid August. They mate in July and lay eggs in small batches on top of the developing florets inside buds (Persson, 1963; Redfern, 1968). The eggs (fig. 4) hatch in 1–2 weeks, the larvae moulting once inside the eggshell. The second-instar larva burrows down a floret to the achene and the receptacle around it swells. Normally, several larvae burrow down separate florets simultaneously and gall tissue forms round each separately. In July and August the gall tissue is soft and the larva feeds on a nutritive zone lining its gall-cell.

instar: stage between moults

It moults into the third (final) instar and the gall tissue outside the nutritive zone becomes woody and increases to coalesce with similar tissue around adjacent cells, forming a hard many-chambered gall (pl. 5.1). The sizes of galls vary between 1 and 34 gall-cells, less than 10 being most common. The larva feeds head downwards and by late August is fully fed, tightly fitting its cell and blocking the entrance with a dark brown posterior spiracular plate (pls. 5.1, 5.2). Fully fed larvae overwinter; in late April and May the larva reverses its position, its skin hardens and darkens to form a puparium (pl. 5.3) typical of higher Diptera and a pupa is formed inside. Adults emerge in June and July (fig. 3).

spiracle: pore where the respiratory system opens at the surface

Adult behaviour

Little is known of the behaviour and causes of death of adult *U. stylata*. Males are common on *C. vulgare* in early June. Females do not appear on plants until July when flower buds are developing. Perhaps they feed

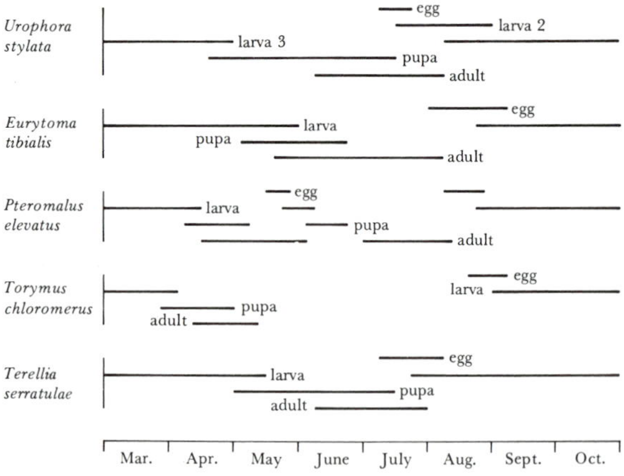

Urophora stylata: egg, larva 2, larva 3, pupa, adult
Eurytoma tibialis: egg, larva, pupa, adult
Pteromalus elevatus: egg, larva, pupa, adult
Torymus chloromerus: egg, larva, pupa, adult
Terellia serratulae: egg, larva, pupa, adult
Mar. Apr. May June July Aug. Sept. Oct.

Fig. 3. Life cycles of *Urophora stylata*, its insect parasites and of *Terellia serratulae* on *Cirsium vulgare*. (*Pteromalus elevatus* and *Torymus chloromerus* probably parasitise other hosts between May and September.)

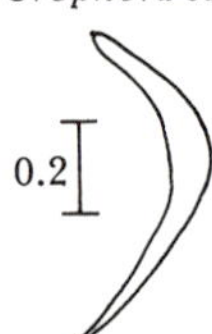

Fig. 4. *Urophora stylata* egg.

during that month enabling their eggs to mature (males may not need to feed). The most likely foods are nectar and aphid honeydew (Varley, 1947) and supplies of these may affect fecundity. It seems that males establish territories on the plant which attract females and provide adequate supplies of buds in which to lay eggs (Zwölfer, 1972). A male patrols a section of plant which he defends against other males of the same and, perhaps, other tephritid species. He faces his antagonist and displays, using scissor movements of the wings and spreading them (no doubt wing pattern is important here: pl. 2.1), meanwhile rocking his body; this usually drives off the intruder, but occasionally they touch and fight, one eventually chasing off the other.

Many questions are unanswered. How big is the territory defended by the male? Perhaps it must contain a minimum of buds of suitable size. Does territory size vary depending on population density of flies? Or does it remain constant with surplus males having to emigrate in order to breed? This could prevent overpopulation by *U. stylata* and may be one mechanism by which population density is regulated.

density: number per unit area (e.g. no./m^2)

Although hybrid matings do occur, especially in captivity, females usually mate with males of their own species. The territorial display of the male probably attracts a female and leads to mating. The sequence of movements and pattern of wings and body of both sexes are probably crucial in species recognition.

Fig. 5. *C. vulgare* bract.

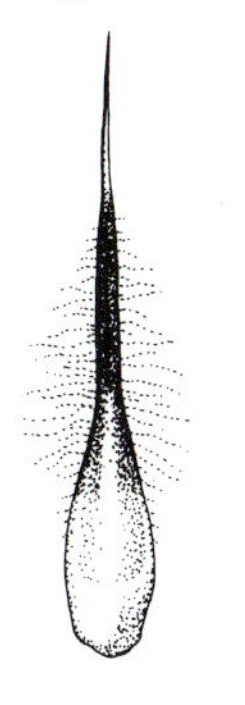

Fig. 6. *C. arvense* bract.

Fig. 7. *U. stylata* puparium after emergence of fly.

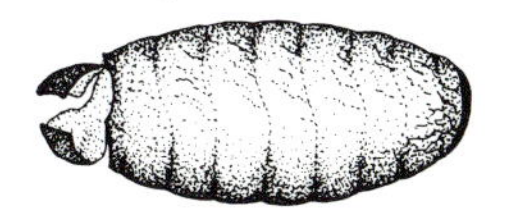

Oviposition

To a female ready to lay eggs, the shape of buds and bracts is the attractive part of the plant (Zwölfer, 1972). She prefers buds of 8–11 mm across, with narrow elongate bracts with spines at least 2 mm long projecting outwards radially (figs. 1 and 5). Compare a bud and bract of *C. arvense* (figs. 2 and 6); their shape may explain why *U. stylata* is so seldom found in this species. Various experiments can be devised to confirm the attractive features of *C. vulgare* buds, using buds of different sizes, with normal bracts, with clipped bracts, or using buds of similar size of other species (*C. arvense*, *Carduus nutans*, *Centaurea*, *Arctium*), or using models of a fragment of bud stuck on a piece of stem or cork. If the bud is attractive, the female will probe it with her ovipositor. She will deposit eggs only if the florets inside are the right size. Hatching of eggs must be synchronised with floret development for successful gall formation.

Mortality factors

There are many causes of death for *U. stylata* larvae (Redfern, 1968; Zwölfer, 1972; Cameron & Redfern, 1974): insect parasites, predators, competition for food

endoparasite: a parasite developing inside its host

ectoparasite: a parasite developing outside its host

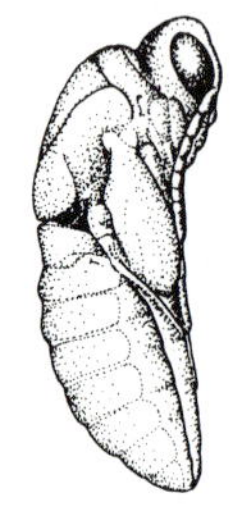

Fig. 8. *Pteromalus elevatus* pupa ♀.

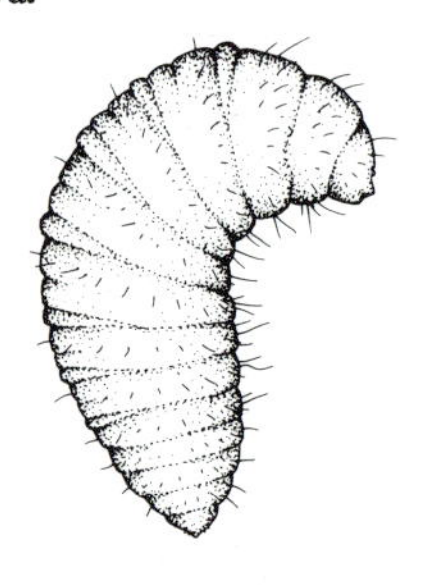

Fig. 9. *Torymus chloromerus* larva.

hyperparasite: a parasite of a parasite

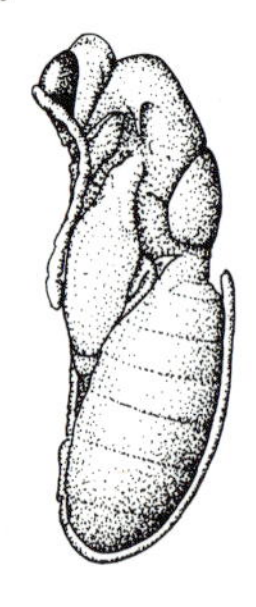

Fig. 10. *T. chloromerus* pupa ♀.

phytophagous: plant-feeding

and space between individuals of the same and different species, disease. The insect parasites known from *U. stylata* are chalcids (Hymenoptera). The commonest is *Eurytoma tibialis* (pls. 7.4, 7.5), an endoparasite. Its eggs are laid inside the second-instar larva of *U. stylata* in July and development is slow until the host is fully fed in August. The parasitised host then reverses its position in its gall-cell and forms a puparium at least 7 months early. Inside, the parasite kills and eats its host before it has formed a pupa. Its life cycle (fig. 3) is well synchronised with its host's. The presence of *E. tibialis* is easy to recognise: before April, unparasitised hosts are still larvae, whereas parasitised hosts have formed puparia. In April and May, puparia can be kept for a month or two and hosts and parasites separated when adults emerge. After emergence, a parasitised puparium has a circular hole in it and contains the yellowish cast skin of the parasite pupa, whereas *U. stylata* emerges through a slit, leaving two flaps behind (fig. 7).

Two ectoparasites are common on *U. stylata*: *Pteromalus* (= *Habrocytus*) *elevatus* (pls. 7.1, 7.2, 5.4; fig. 8) and *Torymus chloromerus* (= *cyanimus* Boheman: pl. 7.3; figs. 9 and 10). Both lay eggs on the third-instar larva of the host in August and their life cycles are less well synchronised with the host than that of *E. tibialis* (fig. 3). Both species have several generations a year, on *U. stylata* and other hosts. Because their eggs are laid after those of *E. tibialis*, they may be found on hosts already parasitised by the endoparasite; they are then hyperparasites of *U. stylata*. If the parasite population is large, they can be hyperparasites via individuals of their own species.

An uncommon insect parasite is *Tetrastichus daira* (fig. 11). It is a gregarious endoparasite, i.e. many individuals are found inside a single host (Redfern, 1972).

The predators of *U. stylata* are more catholic in their feeding habits than the parasites. Larvae of *Palloptera* spp. (Diptera: Pallopteridae) (fig. 12) feed on several insects in thistle heads, including *U. stylata* larvae, curled up inside the prey's empty skin, and caterpillars, found in the frass. The larvae drop from thistle heads in November and burrow into the soil to pupate, and adults (pl. 2.7) emerge in May (Parmenter, 1942; Stubbs, 1969).

The commonest predators of *U. stylata* are caterpillars, especially *Eucosma cana* (pls. 5.7, 5.8) which can cause considerable mortality (e.g. 24%: Redfern, 1968). All species found in flower heads (table 3) are primarily phytophagous. *E. cana* feeds on florets, achenes (fig. 13) and receptacle tissue (see fig. 14 for explanation of floral parts), filling cavities with frass. It is more common in

Fig. 11. *Tetrastichus daira* ♀.

Fig. 12. *Palloptera* sp. larva.

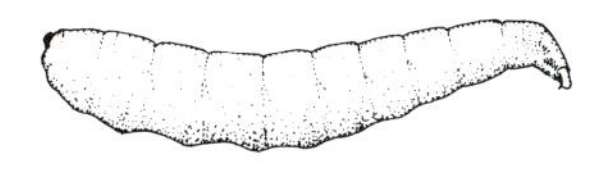

Fig. 13. Caterpillar-damaged achene.

thistle heads with galls than without, presumably attracted by the gall tissue which is succulent in August and September when the caterpillars are feeding actively. Often they break into gall-cells and eat any larvae they meet, leaving cells packed with frass and partially destroyed. In late September the fully fed caterpillars drop to the ground and overwinter in cocoons of silk and soil particles. They pupate between May and July and adults (pl. 3.7) emerge and fly from June to August (Bradley, Tremewan & Smith, 1979). The caterpillars are attacked by insect parasites (ichneumons and chalcids) and predators, probably the most important being insectivorous birds (e.g. goldfinch *Carduelis carduelis*).

Life tables

The effects of different mortalities on a population of *U. stylata* can be compared by analysing yearly samples of galls from the same thistle population, collected immediately after emergence of the gall-flies when causes of death can be recognised (Redfern, 1968). To understand the effects of successive mortality factors on a population throughout a generation, one can construct life tables. This has been done for the Knapweed gall-fly *Urophora jaceana* (Hering) (table 4), though not for *U. stylata*. The distribution of *U. jaceana*, and of other tephritids (Myers & Harris, 1980), is patchy. This is not only due to the distribution of its host plant; distribution is often clumped within uniform stands of the plant.

Life table analysis involves estimating the density of a population at several successive stages of the life cycle (table 4). From this, one can estimate mortality, and

Fig. 14. Floral parts.

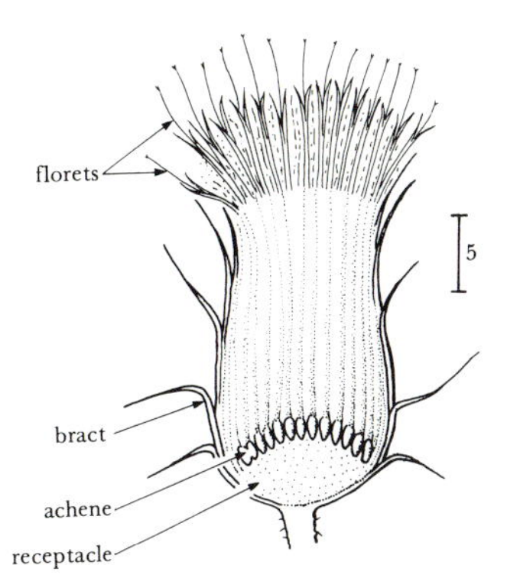

Table 3. *Caterpillars (Lepidoptera) in flower heads of thistles (adult moths are illustrated in pl. 3)*

Species	Main host plants	Larva in head	Pupa	Distribution
Aethes cnicana	*Cirsium*	9–10	In stem	Common, widespread
A. rubigana	*Arctium*	9–10	On ground	Local, widespread
Eucosma cana	*Cirsium, Carduus, Centaurea*	8–9	On ground	Common, widespread
Myelois cribrella	*Cirsium, Carduus*	8–10	In stem	Local, C. & S. England
Homoeosoma nebulella	*Cirsium vulgare, Senecio jacobaea*	8–10	On ground	Local, C. & S. England
Phycitodes binaevella	*Cirsium vulgare, Carduus*	8–9	On ground	Local, widespread

Numbers indicate months.
From Beirne (1954), Bradley *et al.* (1973, 1979).

usually identify its cause, at each stage. Mortality is usually expressed as:

$$\frac{\text{number dying in stage*}}{\text{number entering stage}} \times 100$$

*or number killed by a specific mortality factor if this acts at one particular time and does not overlap with other mortalities.

Some mortalities are very high, e.g. parasitism by *Eurytoma tibialis* and winter disappearance. In winter a large proportion of flower heads (of thistles as well as knapweeds) fall to the ground and disintegrate, releasing their galls. Some are damaged by small mammals which gnaw into individual gall-cells from the base. Others disappear from the quadrats and it is likely that small mammals carry them away, as small piles can be found in their runs.

Population dynamics

If life tables are available for one population over several years, there is a graphical method of analysis (*k*-value analysis) which makes it possible to identify the factor mainly responsible for changes in population density from generation to generation (the 'key factor'). This technique is described in full in Varley, Gradwell & Hassell (1973) and Southwood (1966). Life table studies can also reveal factors which regulate the population, maintaining its particular level of abundance. To do this, a mortality factor must kill a greater proportion at high

Table 4. *Life table for one generation of* Urophora jaceana

		U. jaceana killed	No./m² alive
May–June 1935	*U. jaceana* flies		6.9
	42.5% are ♀♀		2.9
July	Mean no. eggs laid/♀ = 70		203
	Infertile eggs	18.3	184.7
	Larvae died before gall formed	37.1	147.6
	Larvae died, causes unknown	3	144.6
	Parasitism by *Eurytoma tibialis* (= *curta*)	65.8	78.8
Aug.–Sept.	Parasitism by *Pteromalus elevatus* (= *Habrocytus trypetae*)	5.6	
	Eurytoma robusta	4.1	
	Torymus chloromerus (= *cyanimus*)	3.7	
	Tetrastichus sp.	0.6	
	Eaten by caterpillars	14.8	50.0
Winter 1936	Winter disappearance	30.8	19.2
	Eaten by mice and voles	12.2	7.0
	Larvae died, causes unknown	1.8	5.2
Spring	Eaten by birds	0.4	
	Parasitism by *Pteromalus elevatus*	0.7	
	Macroneura vesicularis	0.25	
	Tetrastichus sp.	0.25	3.6
July	Drowned in floods	1.57	2.03
	U. jaceana flies for next generation		2.03

Density includes galls on stems and on ground.
From Varley (1947).

Fig. 15. *Terellia serratulae* larva.

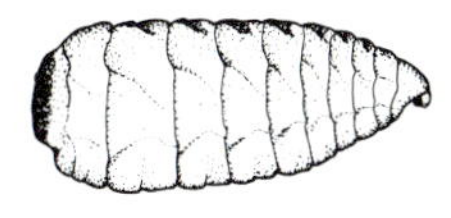

Fig. 16. *T. serratulae* larval cocoons in head.

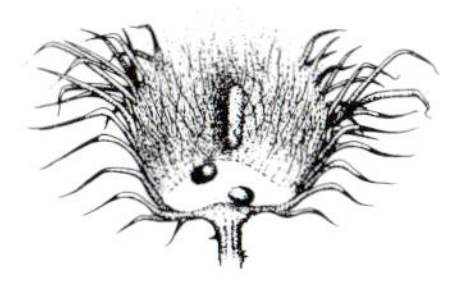

Fig. 17. *Pteromalus musaeus* pupa inside *T. serratulae.*

Fig. 18. *Tetrastichus cirsii* larvae inside *T. serratulae.*

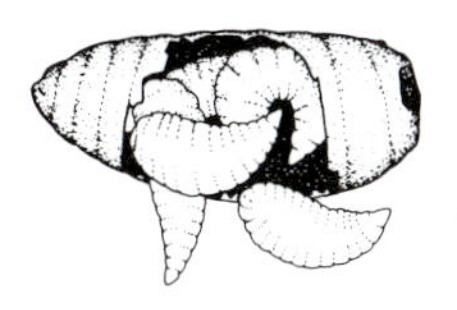

Fig. 19. *T. cirsii* pupa.

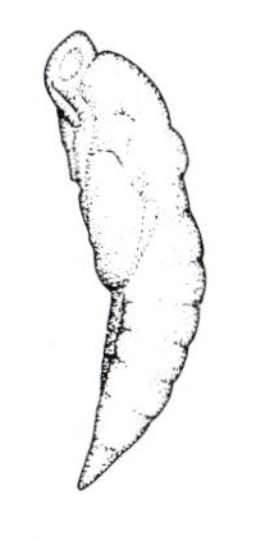

densities of the host population than at low densities: i.e. it must be density dependent (Varley *et al.*, 1973). In *U. jaceana*, possible regulating factors are competition for food and space between young larvae before gall formation, and parasitism by *Eurytoma tibialis* and *Pteromalus elevatus*. To discover which factors are regulating the population needs a long series of generations; different factors may operate at high or low densities and during some generations there may be no regulation occurring. No study of the population dynamics of *U. stylata* has been made; perhaps its populations are regulated in much the same way as those of *U. jaceana.*

Ecological energetics

During the winter, larvae of *U. stylata* (a herbivore, or primary consumer) and its parasites (secondary consumers, representing the next trophic level) sit fully fed in their gall-cells. Their dry weights and energy values (if a bomb calorimeter is available) can be used to illustrate some principles of ecological energetics (see Cameron & Redfern, 1974, for a full description of methods).

Terellia serratulae

Terellia serratulae (pl. 2.8) is a non-gall-inducing tephritid (table 2) and is common in flower heads of *Cirsium vulgare* (and *Carduus nutans* and *C. crispus*) in S. Britain (Redfern, 1968). The larva (fig. 15) constructs a cocoon of silk and floret hairs above the receptacle or partly within it (fig. 16) and feeds on the bases of the florets, achenes and receptacle tissue. There are commonly one to three larvae per head, but occasionally there are up to ten, and most of the receptacle is then eaten. Rarely, larvae are found in heads galled by *U. stylata*; nothing is known about possible competition between these species. The life cycle is similar to that of *U. stylata* (fig. 3).

Causes of death of *T. serratulae* are not well known. The commonest insect parasite is the endoparasite *Pteromalus musaeus* (fig. 17; Redfern, 1968; Graham, 1969). Uncommon are *Tetrastichus cirsii* (figs. 18 and 19) which is gregarious, and *Crataepus marpis* (both Hymenoptera: Eulophidae). Larvae may also be eaten by maggots of *Palloptera* spp. (fig. 12) and moth caterpillars.

Cecidomyiidae (gall-midges)

Small orange cecid larvae are common in flower heads of thistles. Common genera are *Dasineura*, *Clinodiplosis* and *Lestodiplosis*. *Dasineura* larvae are found between bracts, in floret hairs and inside corolla tubes and feed on florets and achenes, sometimes causing distortion or galling of florets and reduction of seed production.

Fig. 20. *Clinodiplosis* sp. larva.

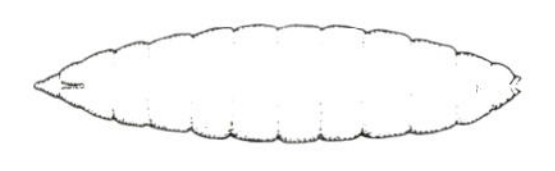

Fig. 21. *Lestodiplosis* larva.

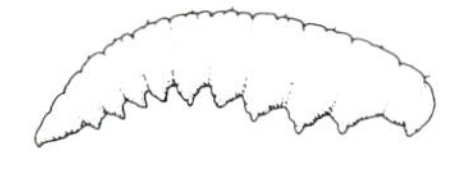

Larvae of *Clinodiplosis* (fig. 20) probably feed on fungi (K. Harris, 1966) and are common in damaged and dehisced heads. *Lestodiplosis* larvae (fig. 21) are predators feeding mainly on phytophagous cecidomyiids (Barnes, 1928).

No detailed study of thistle cecids has been done; host–parasite relationships could prove interesting (see Milne, 1960, on cecids of clover flower heads; Redfern & Cameron, 1978, on a gall-inducing species in yew).

Thistle head community

The community of insects in thistle heads, especially of *Cirsium vulgare* (fig. 22), is large. Thistle heads and plant galls (also *Urophora cardui*, p. 14) form miniature ecosystems in which coexistence and competition between species can be studied (Askew, 1980). There seem to be two strategies for survival for phytophagous species and their insect parasites: *evasion*, which avoids interspecific competition by specialisation of food niche, and *aggression*, which eliminates competitors by direct contact (Zwölfer, 1979). In *U. stylata* galls, the endoparasite *Eurytoma tibialis* attacks first. Later, the ectoparasite *Pteromalus elevatus* will eat host plus endoparasite if it feeds before the host's puparium is formed; it is more aggressive than *E. tibialis* if its timing is right. Like *E. robusta* on *U. cardui* (p. 15) it has a broader host range than *E. tibialis* (equivalent to the endoparasite *E. serratulae* on *U. cardui*) so that it has alternative hosts if populations of *U. stylata* decrease. Theoretically it could eliminate *E. tibialis*, but it is usually less common. In the *U. cardui* system some hosts (those in large galls) are unavailable to *E. robusta*, so providing a food refuge for *E. serratulae* (Zwölfer, 1979). A similar mechanism may exist in the *U. stylata* system and could be tested

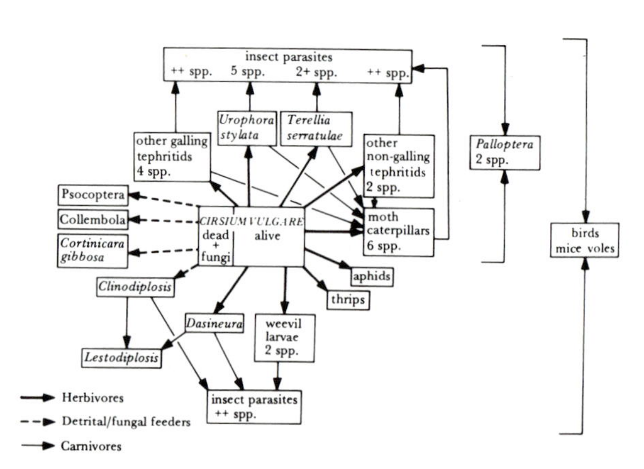

Fig. 22. Simplified food web of insects living in flower heads of *Cirsium vulgare*.

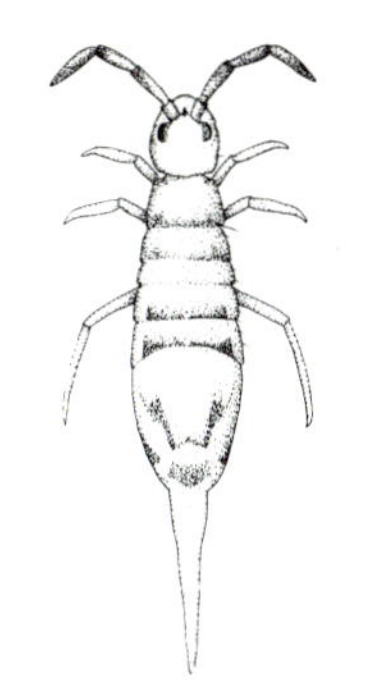
Fig. 23. A collembolan.

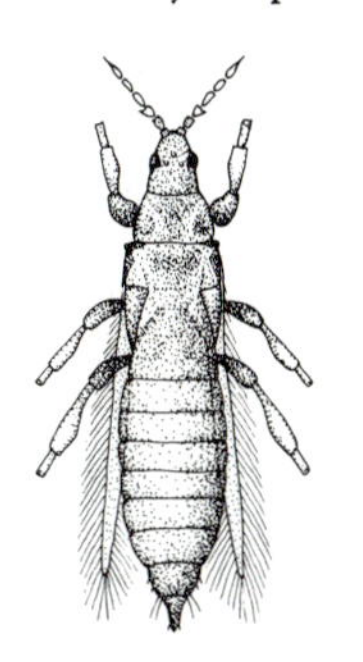
Fig. 24. A thysanopteran.

by examining the percentage parasitism achieved by the two species in galls of different sizes and by recording positions of the parasites in galls.

Insects sheltering in dead heads

By the beginning of winter, thistles are dead and the heads on the plant become dry and brittle, consisting of a receptacle and its hairs enclosed by bracts. The stems and thistle heads may remain standing for several months but, more often, they fall to the ground. The heads become refuges for a different set of animals which probably do not feed. These include springtails (Collembola, fig. 23), thrips (Thysanoptera, fig. 24), froghoppers (Hemiptera: Cercopidae), puparia of leaf-mining flies (Diptera: Agromyzidae), larvae of gall-midges (Diptera: Cecidomyiidae), chalcids (Hymenoptera), beetles (Coleoptera: Carabidae, Staphylinidae, Coccinellidae, Curculionidae), mites, spiders, woodlice, worms and slugs.

3 Stem-borers and leaf-miners

The stems and leaves of thistles support a fauna of flies (Diptera), beetles (Coleoptera) and moths (Lepidoptera) which live as larvae inside the plant, boring or making galls in the stem, mining the leaves, or protecting themselves by folding the leaf or spinning a web underneath which they shelter and feed.

STEM GALL-INDUCERS AND BORERS

Diptera (flies)

The larvae of two species live inside thistle stems: the tephritid gall-fly *Urophora cardui* and the agromyzid stem-borer *Melanagromyza aeneoventris*.

***Urophora cardui* Tephritidae**

Life cycle

The larvae of *Urophora cardui* induce galls in stems of *Cirsium arvense* (pls. 6.1, 6.2, 6.3). The life cycle is similar to that of *U. stylata* (fig. 3). Adults (pl. 2.4) emerge in June and July and females lay batches of eggs into terminal buds. The second-instar larva hatches and burrows into the stem; gall tissue forms round each separately, coalescing and hardening into a many-chambered gall (pls. 6.1, 6.2). There are one to nine cells per gall with an average of three or four. Fully fed larvae overwinter (pl. 6.3), and pupate in the gall, and adults emerge the following summer from galls disintegrated on the ground.

U. cardui is widespread in Europe but has a patchy distribution; it is locally common in S. England but scarce in the Midlands and North (Niblett, 1956). It seems to prefer creeping thistle growing in damp, shady places, though why this should be so is unknown. Other aspects of its life are not well known; adult behaviour, territoriality and courtship are probably similar to *U. stylata*. Recognition of oviposition sites must be different; perhaps size and shape of buds and bracts of *C. arvense* are important as they are in *C. vulgare* for *U. stylata*.

Insect parasites

Four chalcids (Hymenoptera) parasitise *U. cardui*. *Pteromalus elevatus* and *Torymus chloromerus* (p. 8) are less common than *Eurytoma serratulae* (= *tristis* Mayr), specific to *U. cardui*, and *E. robusta*, found on several gall-forming *Urophora* species (Claridge, 1961). The life cycle of *E. serratulae* is similar to that of *E.*

tibialis on *U. stylata* (fig. 3); it is an endoparasite and in late summer it induces the host to form a puparium, in which it overwinters. *E. robusta* is an ectoparasite, common on *U. cardui* and occasionally on *U. stylata* and others (Varley, 1937). Like *P. elevatus* on *U. stylata*, it attacks after *E. serratulae*, laying eggs on third-instar larvae or loose into the gall-cell. It will eat host larva and endoparasite, but if the host succeeds in forming a puparium, *E. robusta* rarely survives (Zwölfer, 1979). It overwinters fully fed (pl. 6.4), pupating in spring and emerging in early summer. *E. robusta* is unusual in that, after eating its host, it starts on the gall tissue lining its cell, leaving behind characteristic chippings of gall and sometimes breaking into adjacent cells where it may demolish other *U. cardui* larvae. It is now acting as an inquiline and, unlike a true parasite, it can grow larger than larvae of its host.

The outcome of competition between *E. serratulae* and *E. robusta* on *U. cardui* varies. *E. robusta* can kill *E. serratulae* and has a wider host range, which means it can survive if the local *U. cardui* population dies out. It can itself cause extinction of a *U. cardui* population, killing such a high proportion that the host colony cannot survive (Zwölfer, 1979). Although inferior in direct competition with *E. robusta*, *E. serratulae* parasitises *U. cardui* in all sizes of galls because it lays eggs in young galls which are small, enabling it to reach all the larvae. Host larvae in the middle of large galls (with more than about three gall-cells) are unavailable to *E. robusta* because its ovipositor is too short; these larvae provide a food refuge for *E. serratulae*, which is sometimes more numerous than *E. robusta*.

E. tibialis is closely related to *E. serratulae* and has very similar habits (Claridge, 1961). It does not attack

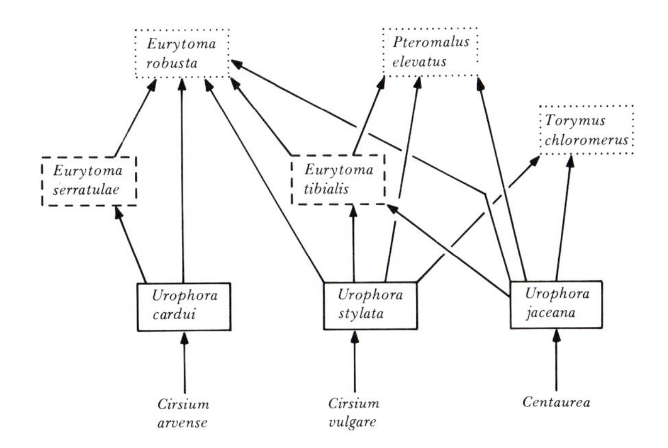

Fig. 25. Host–parasite relationships of three species of *Urophora* (including only the commonest parasite species).

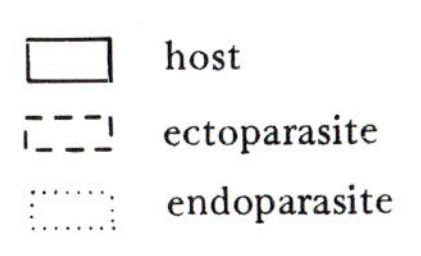

U. cardui, and so its food niche does not overlap with that of *E. serratulae*. *E. robusta* competes directly with both endoparasites, although its place on *U. stylata* is more commonly filled by *Pteromalus elevatus*. *E. serratulae* survives because of its food refuge; perhaps this is also true of *E. tibialis*.

The parasite–host system of *U. cardui* is similar to that of *U. stylata* and *U. jaceana* (summarised in fig. 25), and would be suitable for studies of life tables and long-term population dynamics.

Melanagromyza aeneoventris (= tristis **Rondani***)*
Agromyzidae (mining flies)
Life cycle

The larva of *Melanagromyza aeneoventris* burrows into the stem pith of *Cirsium* spp., less often into Ragwort *Senecio jacobaea*, *Carduus* and *Inula* spp. (Spencer, 1972, 1976). It is widespread in Britain and common in the south, particularly in *Cirsium arvense*. The larva feeds in May and June and pupates soon afterwards. The pale straw-coloured puparium (fig. 26; pls. 5.5, 5.6) is found near the surface of the stem, with an exit hole for the adult prepared nearby, or is separated from the outside by a flimsy skin. Puparia overwinter and flies (fig. 27) emerge in April. Several puparia often occur in one stem (up to six in *C. arvense* from Wyre Forest, Worcs.) and hymenopterous parasites often emerge from them, e.g. *Syntomopus incisus* (Pteromalidae) and an unidentified braconid.

Fig. 26. *Melanagromyza aeneoventris* puparium.

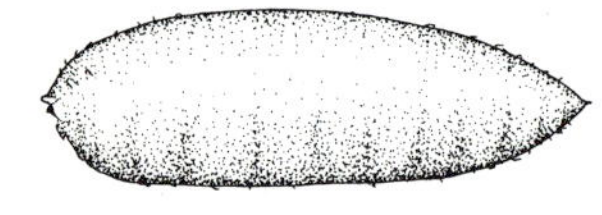

Fig. 27. *M. aeneoventris* adult.

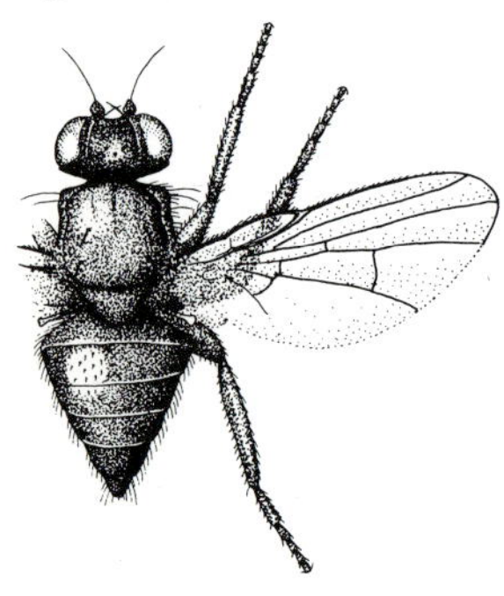

No work has been published on the population dynamics of *M. aeneoventris* and its parasites or on its effects on the thistle and little is known of the habits and behaviour of the adult, despite its being easy to identify and to find in thistle stems.

Coleoptera: Curculionoidea (weevils)

Several species of weevils burrow into thistle stems and roots. Some are rare and occur in S. England only, where they reach the northern limit of their range (see key IV.4); they are more common in continental Europe. *Apion carduorum* (pl. 7.6) and *A. onopordi* are common and widespread south of the Scottish Highlands; *Ceutorhynchus litura* (pl. 7.7) is widespread and locally common (Joy, 1932).

Fig. 28. *Apion* sp. larva.

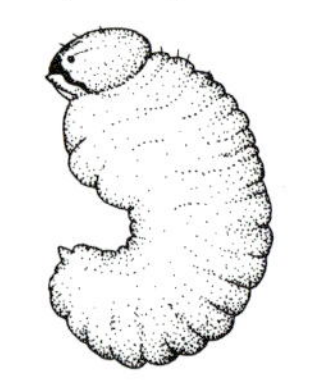

The host plants of these three species vary slightly: *A. carduorum* occurs on *Cirsium vulgare*, *C. arvense* and other thistles and burdocks (*Cirsium*, *Carduus*, *Arctium*); *A. onopordi* on *C. vulgare* but not *C. arvense*, although it also occurs on *Arctium* and *Centaurea* knapweeds; and *C. litura* is almost specific to *C. arvense* (M.G. Morris,

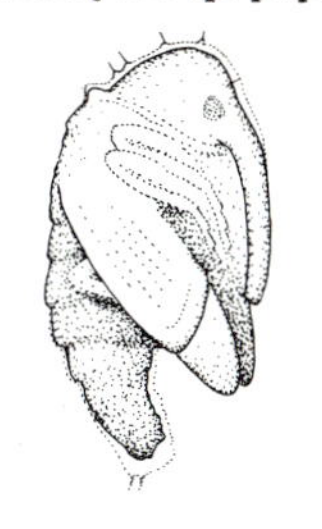

Fig. 29. *Apion* sp. pupa.

personal communication). There is scope here for detailed studies on interspecific competition.

Life cycle

Adults of these species are found on thistle rosettes in April and May, having overwintered in soil and litter. They feed on the leaves, making round holes which usually do not pierce the opposite epidermis. In *C. litura*, eggs are laid in groups of one to five in cavities made by the rostrum on the underside of a leaf, often in the midrib (Zwölfer & Harris, 1966). The hatchlings mine into the stem via the midrib. This becomes black and the leaf dies after a few days. The larva burrows down to the root collar or root, feeding and moulting twice. The third instar leaves the stem in June before the plant is mature and pupates in a cocoon in the soil, and adults emerge in midsummer. They feed on the plants for a short time but by mid-July most have left to find hibernation sites. Egg-laying in *A. carduorum* is similar (Balachowsky, 1963). Larvae of both *Apion* species are found in stem, root collar or root in June and July (figs. 28 and 30) and they pupate (fig. 29) in the larval burrow in July. Adults emerge from mid-July onwards and feed, leaving the plants for overwintering sites in August. Up to nine larvae may be found in a stem but one or two is more common (the two *Apion* species are difficult to distinguish as larvae). In *C. litura*, three to six is common; in a heavy infestation of up to 12 larvae, the individual tunnels fuse and the stem becomes inflated and hardened to form an indistinct gall 3–4 cm long.

rostrum: snout of adult weevil

root collar: swollen base of stem where it joins the root

Mortality factors

Little work has been done on insect parasites and predators. An ectoparasite *Trichomalus gynetelus* (Hymenoptera: Pteromalidae) attacks the *Apion* larvae (personal data) and in Canada a fly maggot *Phaonia trimaculata* (Muscidae) (Zwölfer & Harris, 1966) and a click beetle *Dallopius pallidus* (Peschken & Beecher, 1973) feed on *C. litura*. There must be others and the whole field of population dynamics is awaiting study.

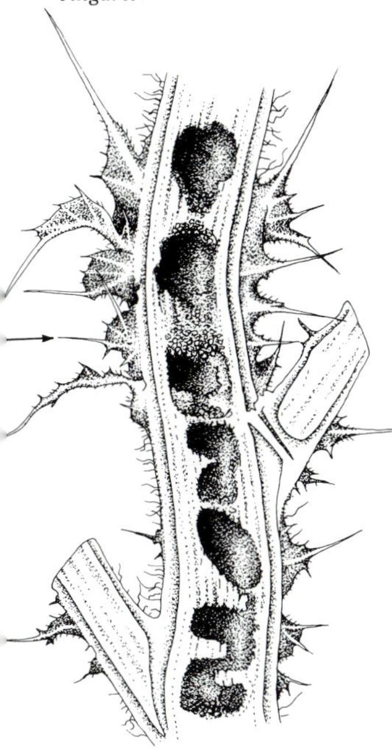

Fig. 30. Damage of *Apion* sp. larva in stem of *Cirsium vulgare*.

Lepidoptera (moths)

Four species of moths are found in *Cirsium vulgare* and *C. arvense* stems (table 5), though none is specific to these plants. No systematic information on parasites is available and good breeding studies would no doubt yield new information on hosts and on parasite life cycles. No quantitative work on population dynamics or effects on growth of the thistles has been done. Competition for stems may occur both between and within species; up to six *Myelois cribrella* larvae may occur in one stem. All species overwinter as larvae in the dead stems, and they usually prepare an exit hole for emergence of the adult.

LEAF-MINERS AND WEB-SPINNERS

These larvae feed on leaves inside a mine or in a web, folded leaf or spun-up shoot. Mines are caused by larvae feeding inside the epidermis or parenchyma of leaves, without breaking through to the outside (fig. 31). The shape and position of the mine help to identify the species within (see key III), but these identifications are tentative only unless the adult has also been named. In thistles, mines are made by larvae of moths, flies and beetles.

parenchyma: green tissue enclosed by upper and lower epidermis of leaf

Fig. 31. Linear mines.

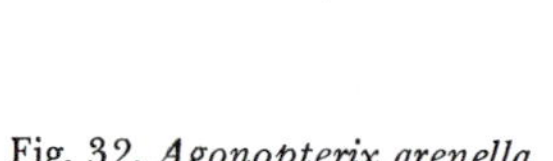

Some caterpillars (Lepidoptera) mine as young larvae, later feeding in spun-up shoots. Others, like *Agonopterix arenella*, spin a web beneath a leaf and feed inside it, usually forming a window in the leaf and not breaking through the upper epidermis (fig. 32). Empty webs of *A. arenella* and other *Agonopterix* species (for adult *A. subpropinquella*, see pl. 3.8) are common on *Cirsium vulgare*.

The biology of thistle leaf-miners and web-spinners is not well known. Good rearing data would add to knowledge of life histories and produce new parasite records. No quantitative work on population dynamics or effects on the host plant has been done.

Fig. 32. *Agonopterix arenella* larva in its web on *C. vulgare*.

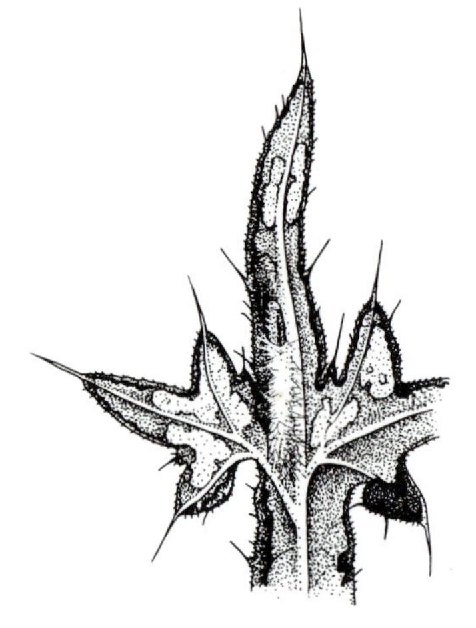

Table 5. *Thistle stem-boring caterpillars (Lepidoptera)*

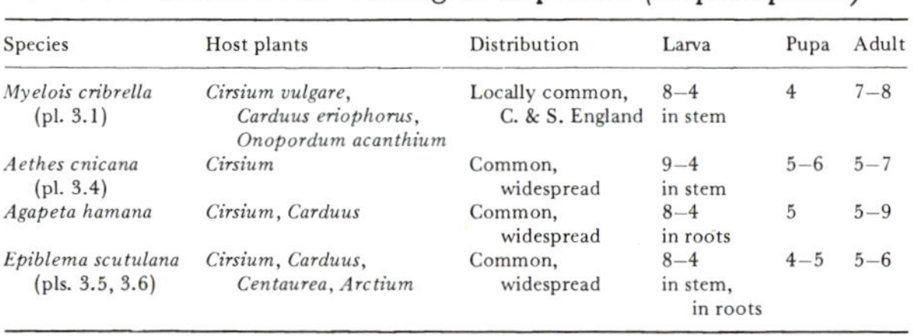

Species	Host plants	Distribution	Larva	Pupa	Adult
Myelois cribrella (pl. 3.1)	*Cirsium vulgare*, *Carduus eriophorus*, *Onopordum acanthium*	Locally common, C. & S. England	8–4 in stem	4	7–8
Aethes cnicana (pl. 3.4)	*Cirsium*	Common, widespread	9–4 in stem	5–6	5–7
Agapeta hamana	*Cirsium*, *Carduus*	Common, widespread	8–4 in roots	5	5–9
Epiblema scutulana (pls. 3.5, 3.6)	*Cirsium*, *Carduus*, *Centaurea*, *Arctium*	Common, widespread	8–4 in stem, in roots	4–5	5–6

Numbers indicate months.
From Beirne (1954), Bradley *et al.* (1973, 1979).

4 Insects on the outside of the plant

A great variety of insects may be found on, rather than in, thistle plants (keys IV.1–IV.4). Some are herbivores feeding on leaves and sap, others visit the flowers for pollen and nectar, while others are casuals, resting or sheltering on the plant and not dependent on it for food.

Herbivorous insects fairly specific to thistles

Hemiptera: Tingidae (thistle lacebugs) (pls. 8.1, 8.2, 8.3)

The two species of lacebugs found on thistles (table 6) have distinct microhabitats: *Tingis ampliata* is almost specific to *Cirsium arvense*, on vegetative buds, stems and leaves; *T. cardui* is usually found on *C. vulgare*, restricted to the bracts of buds and flower heads or the young leaves just below (Southwood & Scudder, 1966; Redfern, 1968). They feed, like aphids, by piercing the plant with fine stylets, tapping sap in phloem vessels.

phloem: plant tissue transporting sugars from leaves

Life cycles

The life cycles of the two species are similar (table 6). Adults overwinter in litter and moss and appear on thistle rosettes in May. *T. ampliata* inserts eggs into the stems of *C. arvense*, while *T. cardui* embeds its eggs in the midrib of young leaves beneath buds of *C. vulgare*. Hatchlings of *T. ampliata* move to vegetative buds and then move very little until the fourth and fifth instars when they, and the new adults in August, are found on mature leaves. *T. cardui* feeds on bracts, sometimes in large colonies; new adults appear in August, when most parent individuals are dead; in late August and September they disperse to overwintering sites.

Distribution

T. cardui is more widespread than *T. ampliata* and spans the range of its host plant. It does not spread on to *C. arvense* in the absence of *T. ampliata*, which suggests that it is not interspecific competition for food that prevents it from colonising *C. arvense*. *T. cardui* flies

Table 6. *Thistle lacebugs* Tingis *spp. (Hemiptera: Tingidae)*

Species	Host plants	Distribution	Egg	Nymphs I–V	Adult
T. cardui (pls. 8.2, 8.3)	*C. vulgare* (rarely *C. palustre* and *Carduus nutans*)	Common, widespread	6–7	7–10	7–7
T. ampliata (pl. 8.1)	*C. arvense* (rarely *C. palustre*)	Common, S. England to Yorks.	5–7	6–9	8–7

Numbers indicate months.
From Southwood & Scudder (1956), Redfern (1968).

actively in May searching for a host plant which, because *C. vulgare* is biennial and reproduces sexually, may be some distance from the previous year's plant; it flies again in September, when dispersing to hibernation sites. This ability to fly no doubt explains its widespread occurrence. *T. ampliata* rarely flies; its distribution is more restricted than its host plant's and it is absent from Ireland and other islands. Its host plant *C. arvense* is perennial and largely vegetative in its reproduction so that the bug will always find new shoots near its overwintering site.

In a field of thistles, the distribution of lacebugs is clumped, some plants having large numbers while others have none. There has been no detailed study of the movements of lacebugs, either from hibernation site to thistle or when on the thistle. Individuals can be marked with paint, which would enable their movements to be followed. In some years, numbers on particular plants are high. Does the population ever reach a point where there is intraspecific competition for food? If so, do the bugs disperse to other plants? Do they have a detrimental effect on the thistle?

Mortality factors

Factors which regulate lacebug populations are unknown. An egg parasite (Hymenoptera: Mymaridae) has been recorded from *T. ampliata* (Southwood & Scudder, 1956) and the ubiquitous predator *Anthocoris nemorum* (Hemiptera) (pls. 4.7, 4.8) feeds on *T. cardui* (Redfern, 1968). Losses during adult dispersal in spring may be the most important mortality factor.

heteroecious: alternating between primary and secondary host plants in one year

autoecious: on one host plant all year

aptera: wingless ♀ (*plural* apterae)

alata: winged ♀ (*plural* alatae)

parthenogenesis: virgin birth

honeydew: sugary liquid exuded from anus

Hemiptera: Aphidoidea (aphids)

Many species of aphids feed on thistles (table 7). Four are common polyphagous species; the rest are restricted to *Cirsium* and its relatives in the Cynareae.

Life cycles

Aphid life cycles are complicated, with alternating phases of sexual and asexual reproduction often involving primary and secondary host plants. The life cycle of *Aphis fabae* is typical of heteroecious species (Blackman, 1974). It infests thistles in early summer and rapidly builds up large numbers of apterae and nymphs in dense colonies, reproducing parthenogenetically. As food supplied by one plant deteriorates, alatae appear and disperse to start colonies on other plants. Sexual forms are found on the primary host plant in autumn and winter. Autoecious species must leave the thistle plant when it dies in autumn; perhaps they overwinter in litter or on rosettes of next year's plants.

Fig. 33. Ladybird larva.

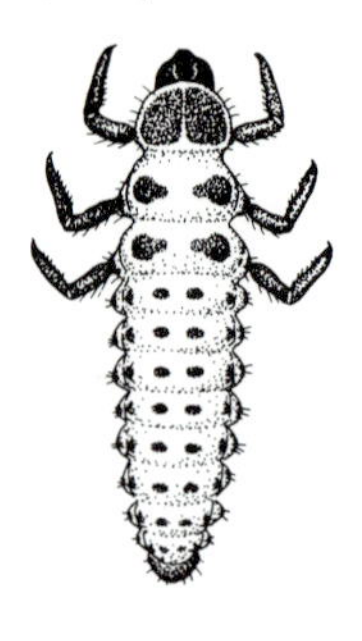

The honeydew produced by aphids often attracts

Fig. 34. *Coccinella 7-punctata* (seven-spot ladybird).

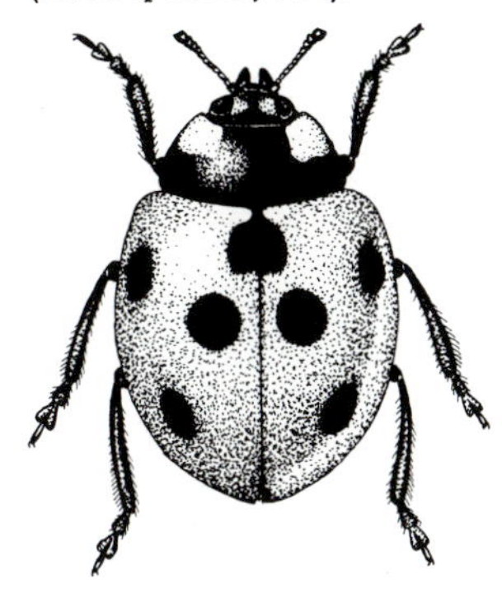

Fig. 35. Hover-fly larva.

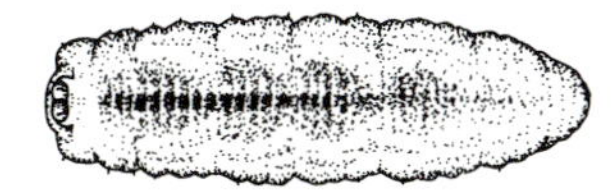

Fig. 36. Aphid mummy.

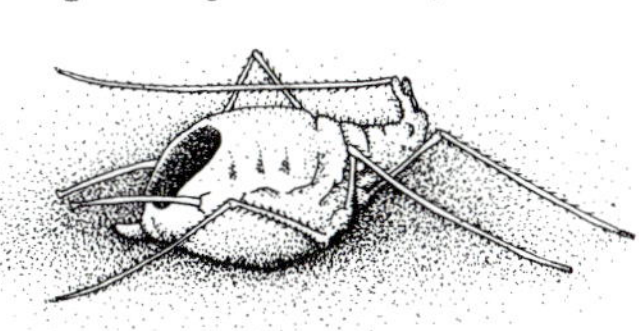

Fig. 37. Aphid mummy caused by *Praon* sp.

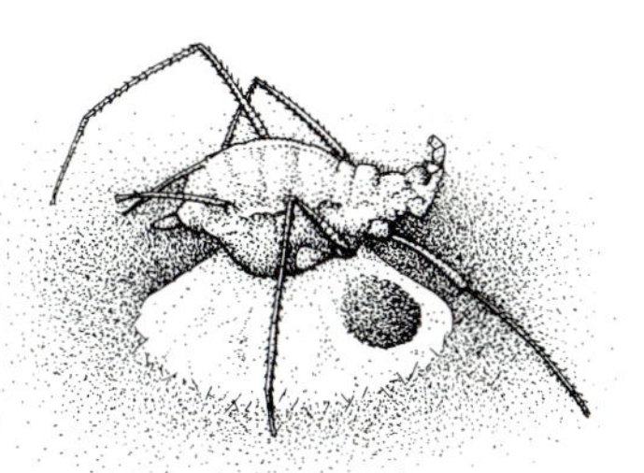

ants. Ant-attended colonies of *A. fabae* grow faster than unattended ones (Banks & Nixon, 1958). Perhaps this is true also of colonial *Dactynotus* and *Brachycaudus* species. Ants also protect aphids from predators, driving off intruders and removing ladybird (Coccinellidae, figs. 33 and 34) and hover-fly (Syrphidae, fig. 35) larvae. Other predators are larvae of lacewings (Neuroptera) and cecidomyiid midges *Aphidoletes* (= *Phaenobremia*) which, despite their small size, can wipe out an aphid colony (K. Harris, 1973).

Most aphid parasites belong to the Aphidiidae and Aphelinidae (Hymenoptera). They are endoparasites and, when about to pupate, cement the aphid corpse to a leaf; this mummy becomes dry and straw-coloured (fig. 36). *Praon* spp. (Aphidiidae) (fig. 37) are easy to recognise, pupating in a cocoon beneath the corpse. The adult parasite emerges through a hole cut in corpse or cocoon. In summer, parasites develop rapidly, perhaps completing several generations on one aphid colony, so that they may help to keep aphid numbers down. However, they seem to be most common in aphid colonies already on the decline and producing alatae to colonise other plants (Blackman, 1974).

Table 7. *Thistle aphids (Hemiptera: Aphidoidea)*

Species	Host plants (secondary hosts)	Feeding site	Life cycle[a] (and primary host)
Protrama radicis	*Cirsium arvense*, *C. oleraceum*,[b] *Arctium*	Roots	Autoecious
Trama troglodytes	*C. arvense*, *Arctium lappa*	Roots	Autoecious
Aphis fabae[c]	Polyphagous including *Cirsium*	Stem, leaves	Heteroecious (*Euonymus europaeus*)
Dysaphis sp.	*C. arvense*, *Arctium*	Roots	Autoecious
Brachycaudus helichrysi and *B. cardui* (pls. 4.1, 4.2)	*Cirsium*, *Carduus*	Leaves, stem, roots	Heteroecious (*Prunus domestica*)
Myzus persicae	Polyphagous including *Cirsium*	Leaves	Mainly heteroecious (*Prunus persica*)
Capitophorus eleagni	*C. arvense*, *Carduus*, *Arctium*	Leaves, stem	Heteroecious (*Hippophae rhamnoides*)
C. carduinus	*C. arvense*, *C. vulgare*, *Carduus*	Leaves, stem	Autoecious
Aulacorthum solani	Polyphagous including *Cirsium*	Leaves, stem	Mainly autoecious
Macrosiphum euphorbiae	Polyphagous including *Cirsium*	Leaves	Heteroecious (*Rosa* spp.)
Dactynotus (= *Uroleucon*) *cirsii* (pl. 4.3)	*C. arvense*, *C. palustre*, *C. oleraceum*	Leaves, stem	Autoecious
D. aeneus	*C. arvense*, *Carduus*	Stem	Autoecious

[a] Autoecious: on one host plant all year. Heteroecious: alternating between primary (tree, shrub) and secondary (herb) host plants in one year.
[b] Not native in Britain.
[c] A group of closely related aphids with *Euonymus europaeus* (spindle tree) as primary host: *A. fabae sensu stricto* may be found on *C. vulgare* as secondary host, *A. fabae cirsiiacanthoides* and *A. fabae cirsiioleracei* on *C. arvense*.
From V.F. Eastop (personal communication), Redfern (1968), Blackman (1974).

Sometimes, predators and parasites must compete for aphid prey. *Aphidoletes* is often the most numerous predator; it is also the smallest and so its effect on the aphids may be less than that of other predators. Detailed studies of the population dynamics of thistle aphids and their enemies have not been done. Intraspecific competition for food, which leads to emigration to other plants, may account for short-term peaks and troughs in aphid population numbers but natural enemies usually contribute to the crash and may, in the long term, regulate the aphid population (see references in chapter 9 of Blackman, 1974; for a project on the effect of ladybirds on aphids, see Wratten & Fry, 1980, exercise 32).

Coleoptera: Chrysomelidae (leaf beetles) (table 8)
Tortoise beetles *Cassida rubiginosa* and *C. vibex* are common on thistles, particularly *Cirsium arvense* (Zwölfer & Eichhorn, 1966).
Life cycles
The life cycles of these species are similar, with one generation a year. Adults overwinter in litter and lay eggs on thistles in May and June. They are laid in batches under a brown scale attached to a leaf. They hatch in late June and larvae are fully fed by August. They eat out windows on the underside of leaves, leaving the upper epidermis intact. They cement their faeces into a mass which, with cast skins, is held by long spines at the posterior end as an umbrella over the larva

Table 8. *Thistle leaf beetles (Coleoptera: Chrysomelidae)*

Species	Host plants	Distribution
Lema cyanella	*C. arvense*	Local, Gt Britain
Oulema melanopa and *O. lichenis*	Gramineae, occasionally *Cirsium*	Common, British Isles
Galeruca tanaceti	Polyphagous including *Cirsium*	Local, England
Apteropeda orbiculata (pl. 4.6)	Polyphagous leaf-miner including *Cirsium*	Common, British Isles
Crepidodera ferruginea (pl. 8.4) and *C. transversa*	Polyphagous including *Cirsium*	Common, British Isles
C. impressa	Polyphagous including *Cirsium*	Rare, coast of Hants., Dorset, Devon
Longitarsus luridus	*C. arvense*	Common, British Isles
Psylliodes chalcomera	Cynareae especially *Carduus nutans*	Local, Gt Britain
Sphaeroderma rubidum and *S. testaceum*	Leaf-miner on Cynareae	Common, British Isles
Cassida rubiginosa (pl. 4.5)	*C. arvense*, other *Cirsium*, *Carduus*	Common, British Isles
C. vibex	*C. arvense*, other *Cirsium*, *Carduus*, *Centaurea*	Local, S. England to Yorks., Ireland
C. viridis	Labiatae, occasionally *Cirsium*	Local, British Isles

From Joy (1932), Linssen (1959), Balachowsky (1963), Zwölfer (1965, 1969).

(pls. 8.5, 8.6). This camouflage may afford some protection from predators; one could compare mortality of larvae with and without their faecal umbrellas. The larvae pupate on the plant in July and August (pls. 8.7, 8.8) and adults emerge in August. They feed on the plant, and by September they have left for overwintering sites.

Numbers of tortoise beetles on a plant can be high, leading perhaps to intraspecific competition between larvae and adults. Perhaps crowded individuals emigrate to other plants; they or their faecal umbrellas could be marked with paint to discover their movements. These beetles are more common on *C. arvense* than *C. vulgare*, perhaps because it is more abundant and occurs in dense stands. Performance may be different on the two plants; this could be tested experimentally (see Wratten & Fry, 1980, exercise 29).

Visitors to flowers

Many insects whose larvae feed elsewhere visit thistle flowers for nectar and pollen; hover-flies (Syrphidae) (pl. 4.10) are particularly common. In warm summers butterflies are frequent and the larvae of one species, the Painted lady *Cynthia cardui* (fig. 38), feed on thistle leaves.

Fig. 38. *Cynthia cardui* adult.

Among Coleoptera, predatory soldier beetles *Rhagonycha fulva* (pl. 4.4) are very common on *Cirsium arvense*, feeding on other flower visitors. Distribution of beetles on the flower heads is not random or systematically spaced as one might expect for competing predators, but clumped (Cameron, 1977). Mating pairs account for some of the clumping, and thistles with many heads in flower are more attractive than those with few. Population size of soldier beetles can be estimated by mark–recapture techniques (Cameron, 1977); they are easy to catch and mark and, because they are normally restricted to the flower heads, one can be confident that all individuals in an area have been found.

Casual visitors

A number of insects (included in keys IV.1–IV.4 as 'others') may be found on the thistle by chance, resting on it because it happens to be there. Particularly common are alate aphids *en route* to new host plants. Sometimes there is a plague of one species which for a short time is found on almost any plant, or tree-feeding aphid species may drop off and land on a thistle beneath.

5 Insect herbivores: biological control

Cirsium vulgare and *C. arvense* are minor crop weeds in Britain but cause serious problems in N. America where they were accidentally introduced. Chemical weedkillers reduce infestations but have disadvantages: they are expensive, their effects are temporary and they may injure the habitat. A more permanent and cheaper solution is a biological agent to control the weeds. In Europe, many insects feed on thistles (Zwölfer, 1965) and help to reduce the size of populations; in N. America, most of these herbivores are missing so that thistle populations can become very large. Several insects are being investigated for introduction into N. America to control thistles and knapweeds (DeBach, 1964; P. Harris, 1980*a*, *b*; Myers & Harris, 1980).

Urophora stylata **Tephritidae**

A promising candidate for biological control of *C. vulgare* is *Urophora stylata* (Zwölfer, 1972). It is almost specific to Spear thistle and unlikely to infest economically useful plants. *C. vulgare* reproduces exclusively by seed, and in galled heads the number of viable achenes is significantly reduced (table 9). Germination success of viable achenes (fig. 39) is the same (70–80%, on moist filter paper in petri dishes) whether from galled or non-galled heads, but all swollen achenes (fig. 40) fail to germinate.

Fig. 39. Healthy achene.

Fig. 40. Swollen achene.

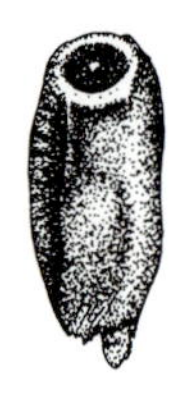

The number of viable achenes increases with size of flower head in galled and non-galled heads (fig. 41). In small heads, galls with three or four cells can reduce viable achenes to nil, while in large heads five or more gall-cells are necessary. Knowing the average number of cells per gall and the density (no./m^2) of galled thistle heads in a population, the effect of *U. stylata* on the

Table 9. *Mean numbers and weights (g) of achenes in galled and non-galled heads of* C. vulgare

	Non-galled heads	Galled heads
No. achenes per head	285.96	141.35
No. viable achenes per head	243.57	47.13
Dry wt achenes per head	0.69	0.17
Dry wt viable achenes per head	0.65	0.10
No. swollen achenes per head	0.00	8.61
(All means significantly different, $P < 0.001$)		

Twenty-three heads in each sample.
From Zwölfer (1972).

seed crop can be calculated. Loss of seed varies between 20 and 70% (Redfern, 1968; Zwölfer, 1972); *U. stylata*, therefore, may be a natural controlling factor in Europe. In N. America without *U. stylata*, *C. vulgare* populations are much larger. If *U. stylata* is introduced successfully, it could become abundant, especially without its European insect parasites, and a balance between gall-fly and spear thistle could result with thistle populations controlled at a low level.

A similar situation exists in flower heads of knapweeds. *Centaurea diffusa* and *C. maculosa*, both of European origin, have become major weeds of grassland in Canada. In 1970 and 1971 two *Urophora* species, both attacking both knapweeds, were introduced to control them (P. Harris, 1980*a*, *b*). Both gall-flies affect the plant by diverting nutrients to galls at the expense

Table 10. *Density of* Centaurea diffusa *and two species of* Urophora *after introduction at one site in Canada*

No. years after release	No. heads/m^2	No. gall-cells/m^2	
		U. affinis	*U. quadrifasciata*
1	2001 ± 263	1.3	0
2	2159 ± 168	300	173
3	3480 ± 185	945	1629
4	1740 ± 447	1444 ± 306	1316 ± 283
5	2382 ± 477	3115 ± 549	796 ± 176
6	1059 ± 12	892 ± 187	431 ± 186
7	614 ± 115	676 ± 145	31 ± 10
8	444 ± 16	276 ± 54	77 ± 17

No. flies released in 1972: 797 *U. affinis* (Frfld.); 25–50 *U. quadrifasciata* (Meigen).
Values are mean ± standard error.
From Harris (1980*a*).

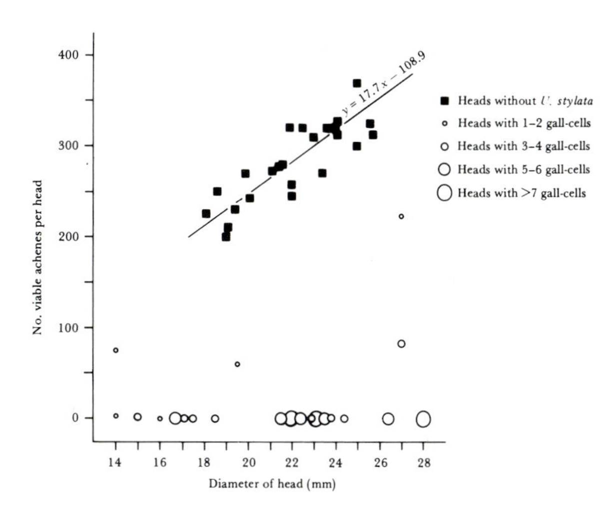

Fig. 41. Number of viable achenes in flower heads of increasing size of *Cirsium vulgare*, with and without galls of *Urophora stylata*. (From Zwölfer, 1972.)

of plant tissues. Number of galls per plant can increase to a critical level; above this, no more nutrients are available for developing galls and surplus *Urophora* eggs and larvae die. It seems, therefore, that *Urophora* density is regulated by the knapweed. On *C. diffusa*, *U. quadrifasciata*, with two generations a year, reached its peak density faster than the single-brooded *U. affinis* and levelled off at a low density (table 10). *U. affinis* may level off at a higher density. Although density of *C. diffusa* declined (and the effects on *C. maculosa* were similar), it has not been reduced to economically acceptable levels. Up to 50% of flower heads were without galls, and as the gall-flies are unlikely to increase in numbers, they cannot eradicate the weed. They have reduced the aggressiveness of the knapweeds, but increased pressure from some other mortality factor would be necessary for complete control.

Urophora cardui **Tephritidae**

Preliminary studies indicate that *U. cardui* reduces the vigour of *Cirsium arvense* (table 11). It is specific to Creeping thistle and may help to control the weed, although it has not been introduced into N. America or elsewhere. Its common insect parasites are absent from N. America and so, like the knapweed gall-flies, it could build up large populations. It may also be worth introducing *U. cardui* to uninfested Creeping thistle in S. England to attempt to control the weed.

Ceutorhynchus litura **Curculionidae**

On *C. arvense* growing under ideal conditions, *C. litura* has no detrimental effect. However, vigour of plants growing in pots in a greenhouse was reduced when they were attacked by more than two larvae per plant (Zwölfer & Harris, 1966). Overcrowded plants containing more than two mature larvae were seriously affected, their leaves decaying and many whole plants dying. Under heavy attack, the plant has a defence: it forms a gall isolating the infestation and allowing the rest of the

Table 11. *The effect of* Urophora cardui *on* Cirsium arvense *in the laboratory*

	Ungalled plants	Plants with 1 gall	Plants with >1 gall
No. plants in sample	31	15	16
Mean no. per plant			
Side shoots	1.9	4.7	4.6
Vegetative shoots	0.6	0.3	0.06
Mean dry weight per plant (g)			
Stems + leaves	3.2	1.7	1.3
Roots	1.3	0.46	0.25

plant to grow normally. This gall does not divert nutrients from elsewhere (unlike the *Urophora–Centaurea* system) and callus tissue can kill the weevil larvae. *C. litura*, therefore, is not a good subject for biological control; it would be effective only on plants growing in poor climatic or overcrowded conditions. Its advantage, however, is that it attacks before flowering and maturation of achenes, so that if conditions were right it could slow down or prevent the spread of Creeping thistle.

No studies have been done on the effects of the commoner weevils *Apion carduorum* and *A. onopordi* on thistles; their potential for biological control may be greater.

Cassida spp. Chrysomelidae

The host plants of the tortoise beetles *Cassida rubiginosa* and *C. vibex* have been investigated (Zwölfer & Eichhorn, 1966). Although both prefer *Cirsium arvense*, laboratory feeding tests showed that they accept a variety of Cynareae and the possibility of their becoming pests of crop plants, such as artichokes, cannot be ruled out. Their introduction into N. America, therefore, is not recommended. In fact, *C. rubiginosa* has been accidentally introduced into E. Canada and the USA where it is quite common on *C. arvense* and, so far, has not spread to other plant hosts (Zwölfer & Eichhorn, 1966). The effect of tortoise beetles on the growth of *C. arvense* is not known but may be marked as both adults and larvae feed on the leaves and populations can be high.

Most work on biological control of thistles has been done with the intention of introducing insects into N. America from Europe. Britain, being a northerly island of Europe, is also missing some of the European thistle fauna, notably the flea beetle *Altica* (= *Haltica*) *carduorum* Guerin. It has been introduced into Glamorgan and Berkshire in an attempt to control *C. arvense* (Claridge *et al.*, 1970). No native species of *Altica* is known to feed on any thistle in Britain and any records of this genus on *C. arvense* or *C. vulgare* would be of interest.

6 Identification

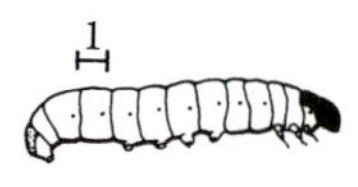

Fig. 42. Caterpillar.

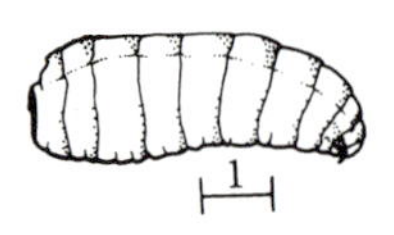

Fig. 43. Maggot.

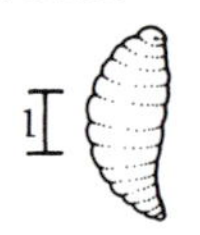

Fig. 44. Grub.

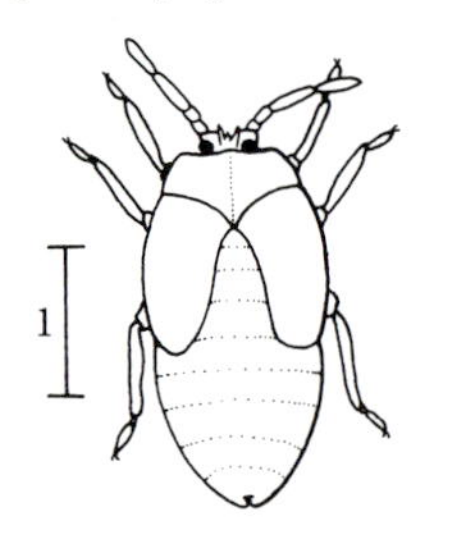

Fig. 45. Nymph.

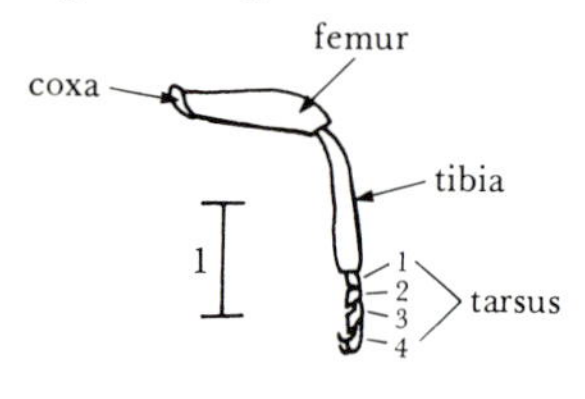

Fig. 46. A leg.

>: more than
<: less than
± : more or less

Introduction to the keys

To use these keys you will have to know exactly where the insect was found. Keys I, II and III deal with those hidden inside the plant, most of which will be immature, and key IV.1–IV.4 with those, mainly adults, living exposed on the outside of the plant.

Immature insects

Many young insects are more difficult to identify to species level than adults. They should be reared to adult and their identification checked in key IV.

The term *larva* strictly applies to the young feeding stages of all insects. Here, it is restricted to the young of endopterygote insects, i.e. those with a pupa in the life cycle and in which the developing wings are hidden inside the body until the pupal stage. Lepidoptera (butterflies and moths), Diptera (flies), Hymenoptera (ants, bees and wasps) and Coleoptera (beetles) are endopterygotes and their larvae are called caterpillars, maggots or grubs (figs. 42, 43 and 44). In exopterygote insects the young are called *nymphs*; in these the wings develop as external buds (fig. 45). In many exopterygotes, nymphs are small versions of the adult, with small wing-buds instead of wings: e.g. Psocoptera (psocids), Thysanoptera (thrips) and Hemiptera (bugs).

Adult insects

To use these keys, adult insects will have to be killed and pinned or preserved in alcohol (see p. 61) and examined under a microscope. Use a stereoscopic microscope magnifying ×25 or more with a good light source and vary the angle of view exhaustively to make sure that you have not missed a character.

You will encounter unfamiliar words when using the keys; some are explained in fig. 46 and the marginal notes and most are illustrated in the diagrams accompanying the keys, which should always be referred to before a character is checked on the specimen. Numbers by scale bars represent millimetres.

Often it is not possible to take an identification down to species, because the insect is immature, or because this can be done only by an expert, or because further identification of that group is outside the scope of this book.

In all keys, references are given to enable you to check your identifications or to identify groups not taken to species level here. These should be consulted if you intend to publish work. The most useful reference is given in bold type.

I Arthropods inside thistle heads (buds, flowering and dead heads)

I.1

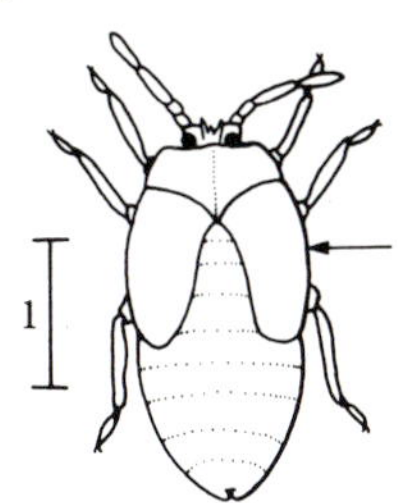

1 With wings. INSECTA — 2

– Without wings or with wing buds (I.1) — 4

2 Adult beetle with first pair of wings hardened (I.2), second pair membranous — COLEOPTERA
(*Cortinicara gibbosa* Herbst. (LATHRIDIIDAE, I.2) is common in flowers in summer. For further identification see key IV.4; Fowler, 1887–91; Joy, 1932; Crowson, 1956; Chinery, 1976.)

I.2

– Both pairs of wings membranous — 3

3 Body squat, antennae at least as long as body, wings held roof-like over abdomen (I.3); venation as in I.4 — (psocids) PSOCOPTERA

– Body slender, antennae shorter than body, wings held flat over abdomen (I.5), fringed with hairs (I.6) — (thrips) THYSANOPTERA

I.3

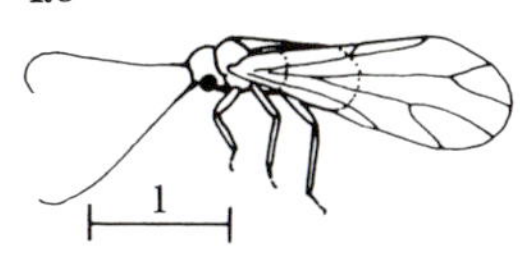

4 With 4 pairs jointed thoracic legs. ARACHNIDA — 5

– With 3 pairs jointed thoracic legs, or none. INSECTA — 6

5 <2 mm long; head + thorax not separated from abdomen (I.7) — (mites) ACARI

– Usually >2 mm long; head + thorax separated from abdomen by waist (I.8) — (spiders) ARANEAE

I.4

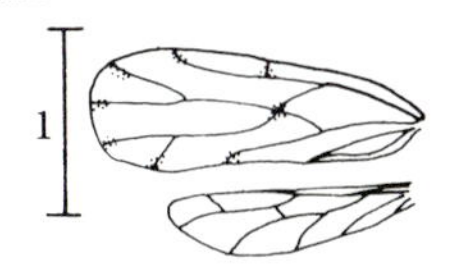

6 With springing organ (I.9); jumps actively — (springtails) COLLEMBOLA

– Without springing organ — 7

7 Mouthparts a piercing proboscis (I.10, I.11); nymph — 8

– Mouthparts not a proboscis; larva (caterpillar, maggot (I.12) or grub), pupa (I.13) or puparium (I.14) — 10

I.5

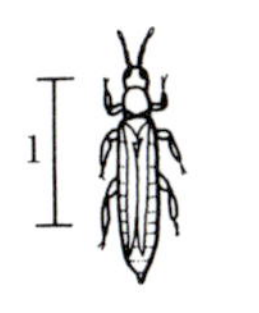

8 Shape elongate and flattened (I.15) — (thrips) THYSANOPTERA

– Not as in I.15. HEMIPTERA (bugs) — 9

I.6

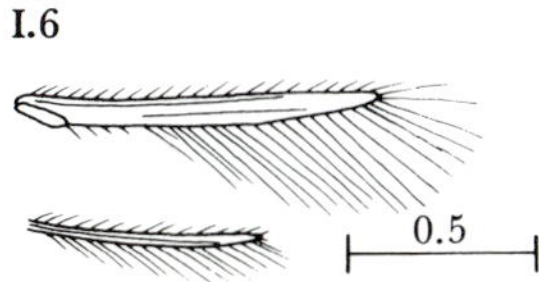

I.7

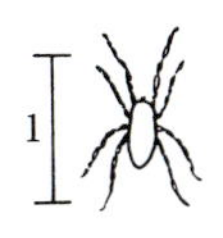

I.8

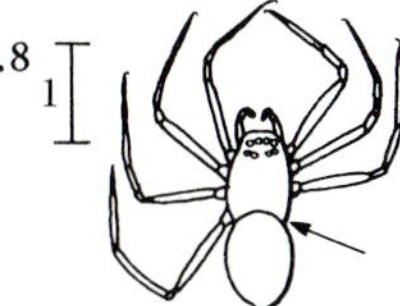

I.9

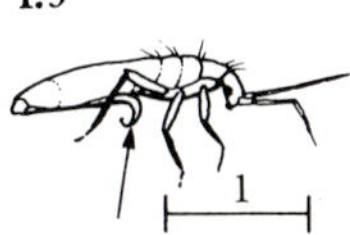

I.10

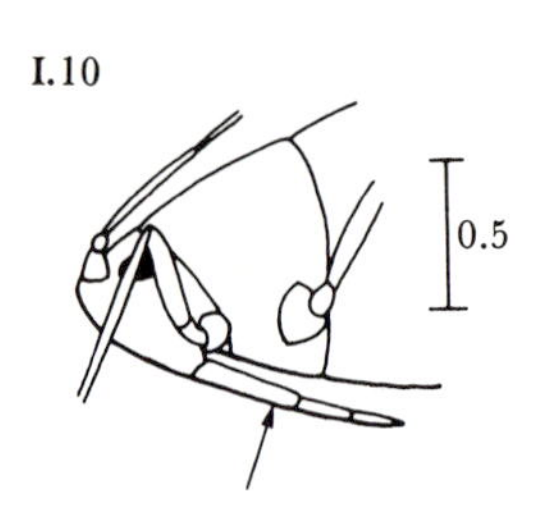

9 With siphunculi on abdomen (I.16); body not flattened. HOMOPTERA (aphids) APHIDOIDEA

– Without siphunculi; body flattened (I.1); usually in *Cirsium vulgare*. HETEROPTERA: TINGIDAE (lace-bugs) *Tingis cardui* (L.)
(Aphids and lacebugs normally are found on outside of plant (see key IV.1); the youngest instars are sometimes found in the heads.)

10 Larva (I.12, I.17) 11

– Pupa (I.13) or puparium (I.14) 29

I.11

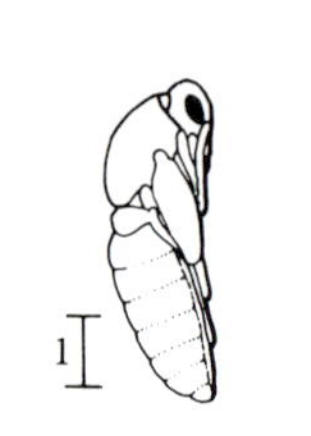

11 Caterpillar; with 3 pairs of legs on thorax and prolegs on abdomen (I.17). LEPIDOPTERA (moths) 12
(Caterpillars which do not fit couplets 12–16 are 'other Lepidoptera': for further identification, see Buckler, 1890–9; Meyrick, 1928; Beirne, 1954; Bradley *et al.*, 1973; Emmet, 1979.)

– Maggot or grub: no legs (I.12) 17

I.12

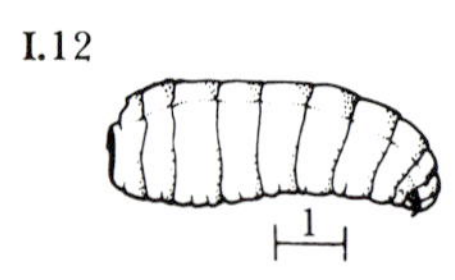

12 Body ± parallel-sided; abdomen without stripes 13

– Body widest at the middle, tapering towards each end (I.18); abdomen with longitudinal stripes. PYRALIDAE 15

I.13

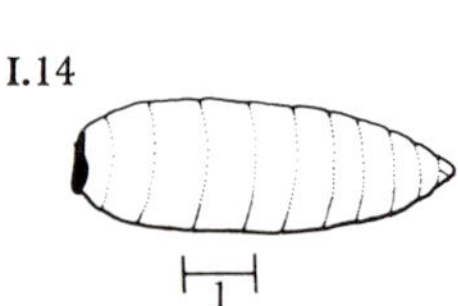

13 Head and prothoracic plate dark brown, abdomen dull pinkish brown; common, in head Aug.–Sept. causing characteristic damage (pl. 5.8). TORTRICIDAE *Eucosma cana* (Haworth)

– Head and prothoracic plate usually light brown, abdomen pale yellow or beige with greenish tinge. COCHYLIDAE 14

I.14

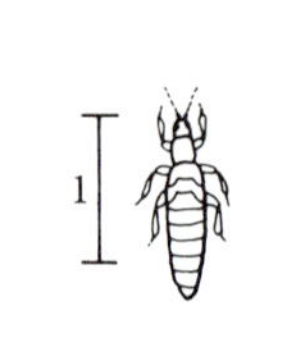

14 Abdomen with conspicuous brown dots (pinacula, I.19); larva in head or stem Sept.–Apr., usually in thistles *Aethes cnicana* (Westwood)

– Pinacula less conspicuous, same colour as abdomen or light grey-brown in full-grown larva; larva in head Sept.–Oct., rarely in thistles *Aethes rubigana* (Treitschke)

I.15

15 Head and prothoracic plate shiny black, abdomen olive green with whitish stripes *Myelois cribrella* (Hübner)

– Head brown, abdomen pale or yellow green with purple or pink stripes 16

I.16

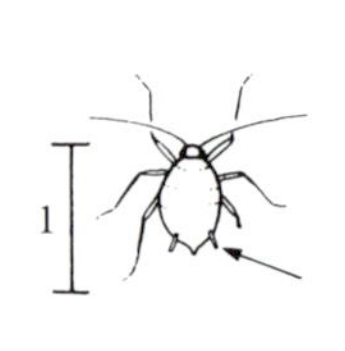

I.17 I.18

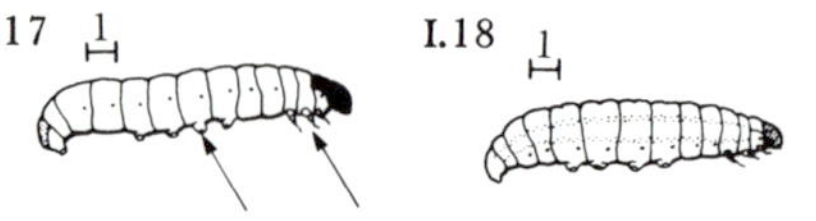

I.19

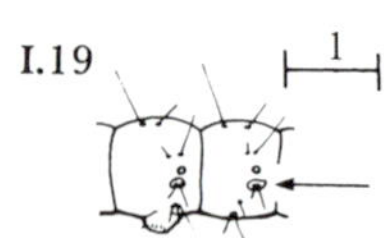

I.20

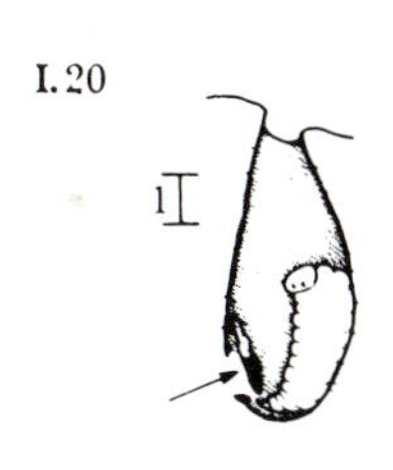

16 Prothoracic plate green, abdomen greenish yellow with dull purple stripes *Homoeosoma nebulella* (D. & S.)

– Prothoracic plate reddish brown, abdomen whitish green with pink stripes *Phycitodes binaevella* (Hübner)

17 Larva not parasitic in or on another larva 18

– Larva parasitic in or on another larva, sometimes in cocoon or gall-cell with shrivelled host remains (I.20). HYMENOPTERA 22

(Couplets 22–27 are intended as a rough guide only and are not comprehensive due to lack of knowledge; all parasites should be reared to adult and, if work is to be published, identifications must be checked by an expert.)

I.21

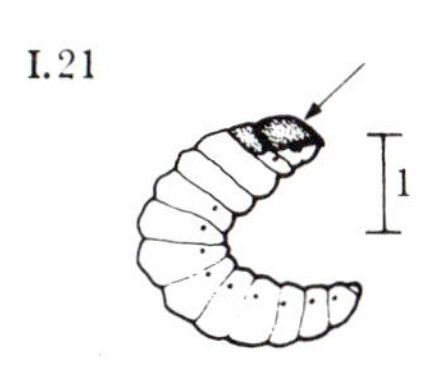

18 With a distinct head (I.21). COLEOPTERA: CURCULIONOIDEA (weevils) CURCULIONIDAE

(*Larinus planus* (F.) and *Rhinocyllus conicus* (Froelich) occur in heads as larvae; both are rare southern species.)

– Without a distinct head (I.22, I.23). DIPTERA (flies) 19

I.22

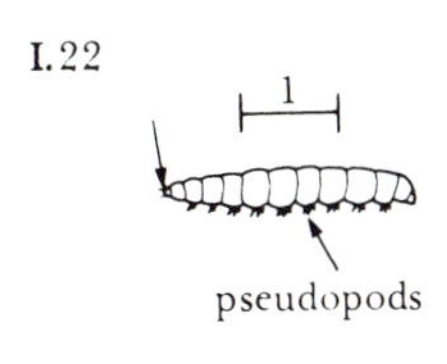

19 Usually 3 mm long or less when fully grown; orange, red or yellow. CECIDOMYIIDAE (gall-midges) 20

– Usually >3 mm long when fully grown; white, dirty white or cream (picture-winged flies) 21

I.23

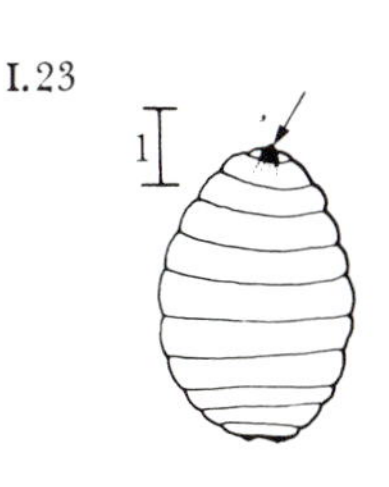

20 With pseudopods, no sternal spatula (I.22) *Lestodiplosis* sp.

– Without pseudopods, with sternal spatula in fully grown larva (I.24) other CECIDOMYIIDAE

(*Dasineura* and *Clinodiplosis* spp. are common.)

21 Elongate, tapering towards anterior end (I.25); may be in cocoon or gall-cell with remains of tephritid larva. PALLOPTERIDAE *Palloptera* sp.

– Barrel-shaped (I.23); in a gall-cell or cocoon of floret hairs TEPHRITIDAE

(Tephritid larvae are commoner in heads of *Cirsium vulgare* than *C. arvense*; *Urophora stylata* (F.) (I.23) is very common in a gall, *Terellia serratulae* (L.) (I.12) in a cocoon of floret hairs.)

I.24

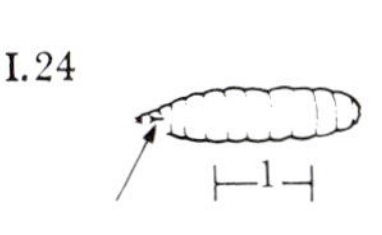

I.25

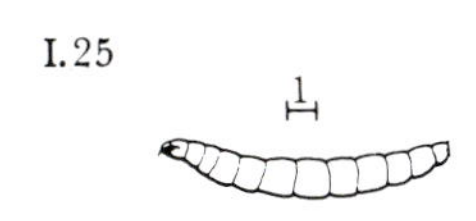

22 Parasite usually of caterpillar; larva often constructs cocoon in which to pupate ICHNEUMONOIDEA

– Parasite usually of Diptera larvae (e.g. Cecidomyiidae); no cocoon PROCTOTRUPOIDEA

– Parasite of larval tephritid or weevil or caterpillar; no cocoon. CHALCIDOIDEA 23

I.26

I.27

I.28

I.29

I.30

I.31

I.32

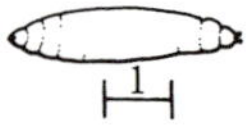

I.33

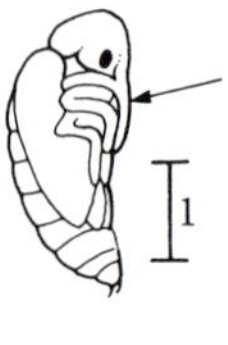

23 Parasite of weevil larva or caterpillar
other CHALCIDOIDEA

– Parasite of tephritid larva 24

24 Gregarious internal parasites of *Urophora stylata* or *Terellia serratulae* (I.26). EULOPHIDAE
Tetrastichus sp.

– Solitary internal or external parasite 25

25 Internal parasite, often inside puparium (hard dark larval skin) of host 26

– External parasite of gall-forming tephritid, inside gall-cell with shrivelled host remains (I.20) 27

26 Inside *Urophora* sp. in gall-cell. EURYTOMIDAE
Eurytoma tibialis Boheman

– Inside *Terellia serratulae* in cocoon of floret hairs.
PTEROMALIDAE *Pteromalus musaeus* Walker

27 Larva active, with long hairs on each segment (I.27).
TORYMIDAE *Torymus chloromerus* (Walker)

– Larva sluggish, not so hairy 28

28 With large mandibles and several long hairs at head end (I.28); egg brown with a tail at both ends (I.29; eggshell often obvious in gall-cell, tails may break off); uncommon in thistle heads. EURYTOMIDAE
Eurytoma robusta Mayr

– With small mandibles and ± hairless (I.30); egg white with papillae, rounded at both ends (I.31); common.
PTEROMALIDAE *Pteromalus elevatus* (Walker)

29 Pupa inside a puparium (segmented, I.14, I.32).
DIPTERA CYCLORRHAPHA

– Pupa free with obvious appendages (I.33), or in a silken cocoon; if in a brown cocoon, it is not segmented 30

30 With elongate snout (I.33). COLEOPTERA:
CURCULIONOIDEA (weevils) CURCULIONIDAE

– Parasite pupa on or in host remains, without elongate snout (I.13) HYMENOPTERA

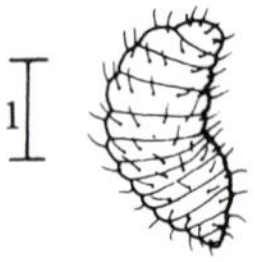

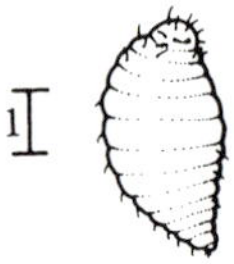

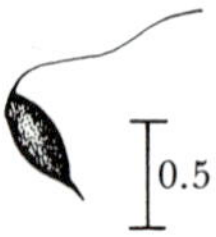

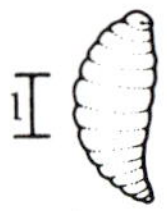

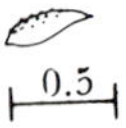

PLATE 1

Cirsium arvense
Creeping thistle (left)
Cirsium vulgare
Spear thistle (right)

PLATE 2

Picture-winged flies (Diptera) whose larvae live in thistle heads, stems or leaves

1
Urophora stylata
Tephritidae ♂

2
Urophora stylata ♀

3
Urophora solstitialis ♀

4
Urophora cardui ♀

5
Xyphosia miliaria
Tephritidae ♂

6
Trypeta zoe
Tephritidae ♀

7
Palloptera umbellatarum
Pallopteridae ♀

8
Terellia serratulae
Tephritidae ♀

PLATE 3

Moths (Lepidoptera) whose larvae live in thistle heads, stems or leaves

1
Myelois cribrella
Pyralidae

2
Coleophora peribenanderi
Coleophoridae

3
Scrobipalpa acuminatella
Gelechiidae

4
Aethes cnicana
Cochylidae

5
Epiblema scutulana
Tortricidae ♂

6
Epiblema scutulana ♀

7
Eucosma cana
Tortricidae

8
Agonopterix subpropinquella
Oecophoridae

9
Phycitodes binaevella
Pyralidae

PLATE 4

Insects living on outside of thistle or visiting flowers (Hemiptera, Coleoptera and Diptera)

1
Brachycaudus cardui
Aphididae, aptera

2
Brachycaudus cardui
Aphididae, juvenile

3
Dactynotus cirsii
Aphididae, aptera

4
Rhagonycha fulva
Cantharidae

5
Cassida rubiginosa
Chrysomelidae

6
Apteropeda orbiculata
Chrysomelidae

7
Anthocoris nemorum
Cimicidae, adult

8
Anthocoris nemorum
nymph

9
Calocoris norvegicus
Miridae

10
Hover-fly
Syrphidae

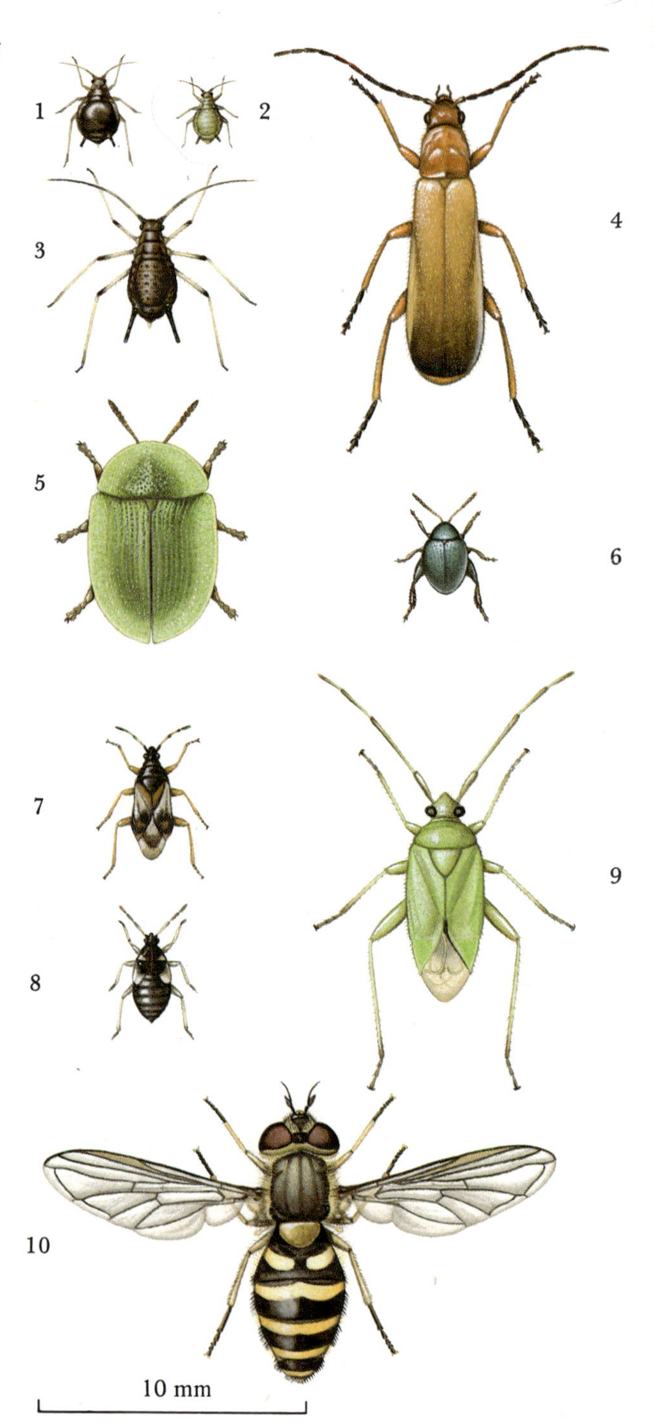

PLATE 5

Insects inside heads and stems of *Cirsium vulgare*

1
Head of *C. vulgare* to show gall-cells of *Urophora stylata* containing (from left to right) *U. stylata* larva; *U. stylata* puparium; *Pteromalus elevatus* larva with shrivelled skin of *U. stylata*; caterpillar frass

2
Urophora stylata fully fed third-instar larva (head towards left)

3
U. stylata puparium (head towards left)

4
Pteromalus elevatus fully fed larva (head towards right)

5
Melanagromyza aeneoventris puparium in stem of *C. vulgare*

6
M. aeneoventris puparium

7
Head of *C. vulgare* cut to show caterpillar and damage of *Eucosma cana*

8
Eucosma cana larva

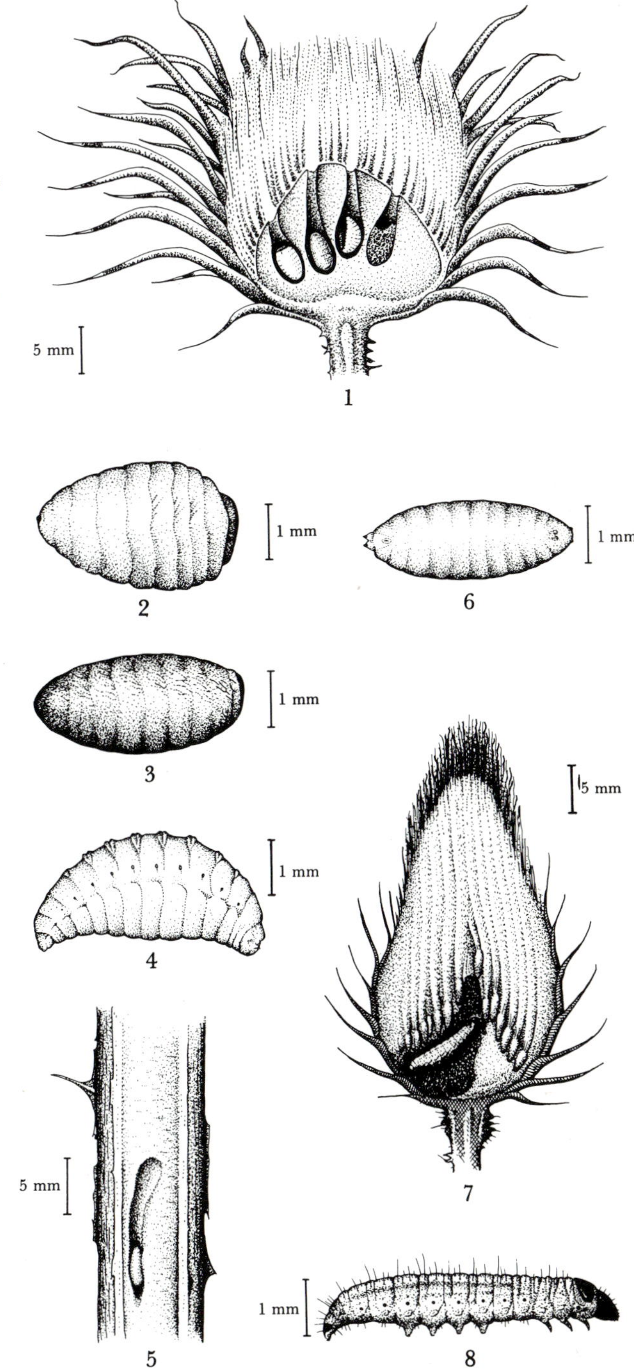

PLATE 6

The gall of *Urophora cardui* in the stem of *Cirsium arvense* and its inhabitants
1
The gall entire
2
Gall cut to show two gall-cells containing *U. cardui* larvae and one gall-cell (central) containing *Eurytoma robusta* larva
3
U. cardui fully fed larva
4
E. robusta fully fed larva

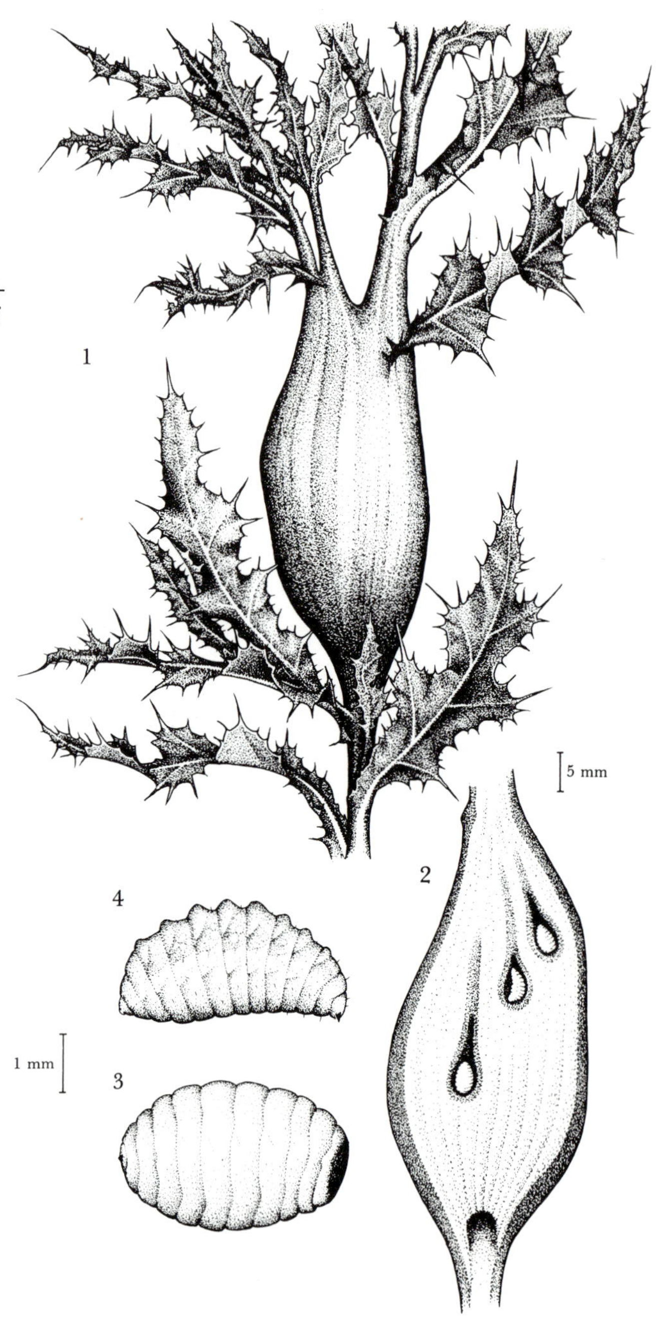

PLATE 7

1–5: Parasitoids (Hymenoptera: Chalcidoidea) of *Urophora stylata*
1
Pteromalus elevatus ♂
2
P. elevatus ♀
3
Torymus chloromerus ♀
4
Eurytoma tibialis ♂
5
E. tibialis ♀
6–8: Beetles (Coleoptera) which feed as larvae in stems or leaves of thistles
6
Apion carduorum
7
Ceutorhynchus litura
8
Sphaeroderma testaceum

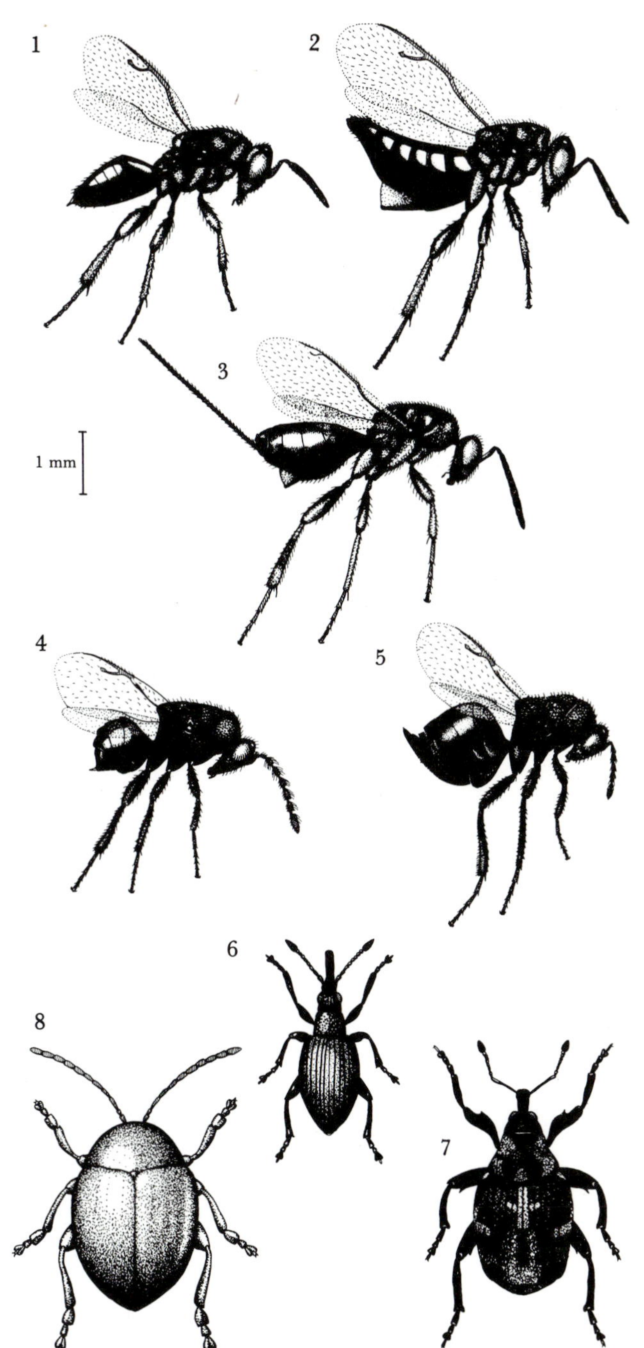

PLATE 8

Bugs (Hemiptera) and beetles (Coleoptera) which feed on the outside of thistle plants
1
Tingis ampliata adult
2
Tingis cardui adult
3
T. cardui nymph
4
Crepidodera ferruginea adult
5
Cassida rubiginosa larva, dorsal view
6
C. rubiginosa larva, side view
7
C. rubiginosa pupa, dorsal view
8
C. rubiginosa pupa, ventral view

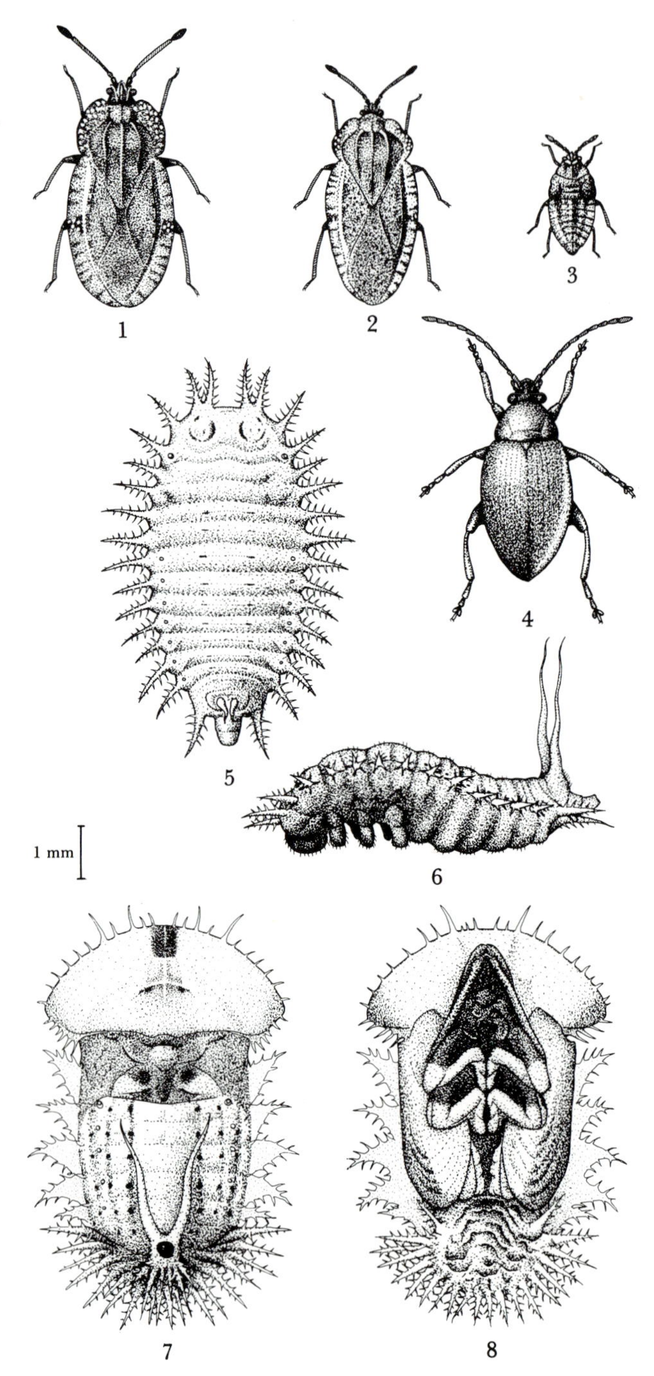

II Insect larvae and pupae inside stems and roots

II.1

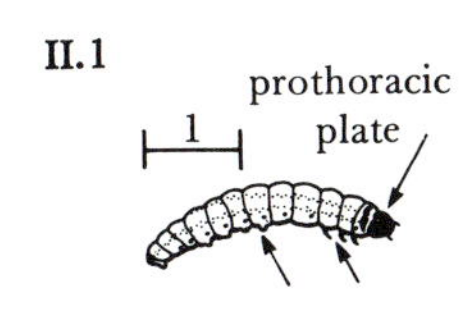

II.2

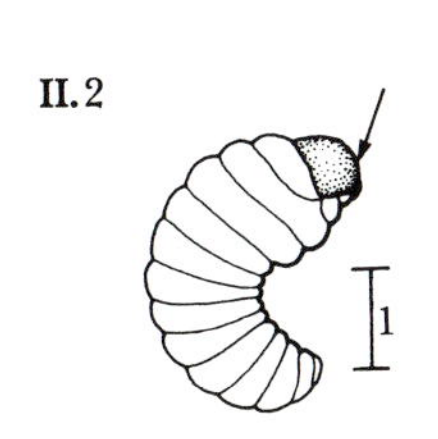

II.3

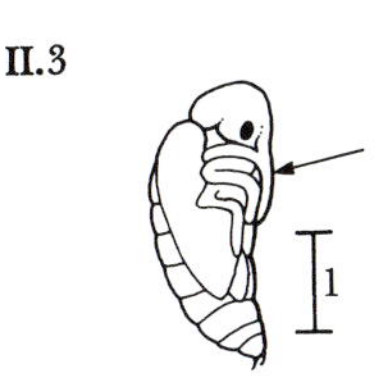

II.4

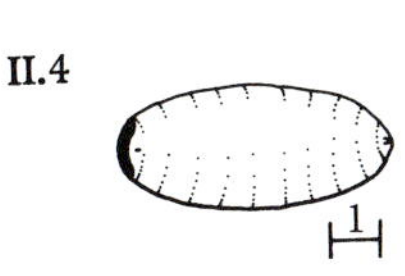

II.5

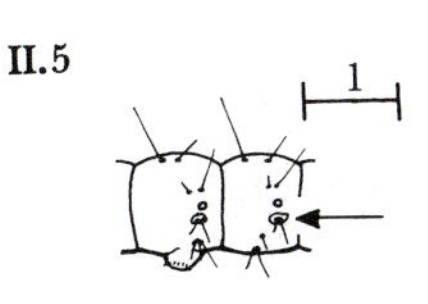

II.6

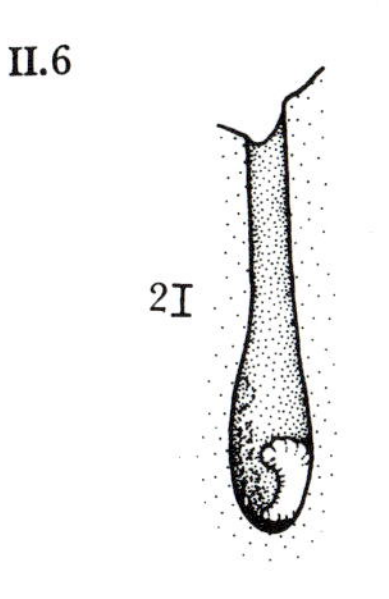

1 Larva: caterpillar (II.1), maggot or grub (II.2) — 2

– Pupa (II.3) or puparium (II.4) — 16

2 Caterpillar: with thoracic legs and prolegs on abdomen (II.1). LEPIDOPTERA — 3

(Two beetle larvae with thoracic legs but no prolegs have been recorded from thistle stems: *Agapanthia villosoviridescens* (Degeer) (CERAMBYCIDAE) and *Mordellistena pumila* (Gyllenhal) (MORDELLIDAE); both are rare from S. England only.)

– Maggot or grub: without legs (II.2) — 6

3 Head and prothoracic plate shining black (II.1); often several larvae in one stem; locally common in S. England. PYRALIDAE *Myelois cribrella* (Hübner)

– Head and prothoracic plate pale or dark brown; usually 1 larva per stem or root — 4

4 Head red-brown, abdomen grey-white, in roots Aug.–Apr. (inside silk cocoon Oct.–Apr.); widespread. COCHYLIDAE *Agapeta hamana* (L.)

– Head light or yellow-brown, abdomen yellow-brown or pink, with brown dots (pinacula, II.5); in stem Aug.–Apr. (rarely in roots) — 5

5 Abdomen pale pink, thoracic legs yellow-brown; common, occasionally in roots. TORTRICIDAE *Epiblema scutulana* (D. & S.)

– Abdomen pale yellow-brown with greenish tinge, thoracic legs dark brown; common, not in roots. COCHYLIDAE *Aethes cnicana* (Westwood)

(Caterpillars which do not fit couplets 3–5 are 'other Lepidoptera': for further identification, see **Buckler, 1890–9**; Meyrick, 1928; Beirne, 1954; Bradley *et al.*, 1973, 1979; Emmet, 1979.)

6 Larva not parasitic on another larva — 7

– Larva parasitic on or in a host larva, sometimes in puparium, cocoon or gall-cell with shrivelled host remains (II.6). HYMENOPTERA — 10

(Couplets 10–15 are intended as a rough guide only and are not comprehensive due to lack of knowledge; all parasites should be reared to adult and, if work is to be published, identifications must be checked by an expert.)

II.7

mouth hooks

7 With a distinct head (II.2). COLEOPTERA (weevils) CURCULIONOIDEA

(*Apion carduorum* Kirby (II.2) and *A. onopordi* Kirby (APIONIDAE) are common in thistle stems.)

– Without a distinct head (II.7). DIPTERA 8

II.8

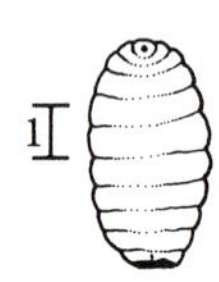

8 Larva barrel-shaped (II.8); in gall in stem of *Cirsium arvense* (pls. 6.1, 6.2, 6.3). TEPHRITIDAE *Urophora cardui* (L.)

– Larva not barrel-shaped; not in a gall 9

II.9

9 Anterior spiracles close together on dorsal surface of prothorax, posterior spiracles on projections on last segment (II.7); larva inactive feeding on stem pith AGROMYZIDAE

(The commonest stem-boring agromyzid is *Melanagromyza aeneoventris* (Fallèn).)

– Larva not as in II.7; herbivore, scavenger or predator other DIPTERA

II.10

10 Parasite usually of caterpillar; often constructs a cocoon in which to pupate ICHNEUMONOIDEA

– Parasite of larval tephritid, agromyzid, weevil or caterpillar; no cocoon. CHALCIDOIDEA 11

11 Parasite of *Urophora cardui* (Tephritidae) in gall-cell 12

– Parasite of agromyzid, weevil or caterpillar 15

II.11

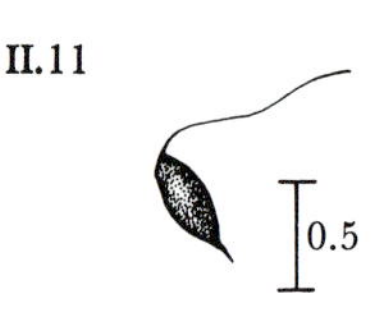

12 Internal parasite, inside puparium of *U. cardui* for most of its life (II.4). EURYTOMIDAE *Eurytoma serratulae* (F.)

– External parasite, with shrivelled host remains (II.6) 13

II.12

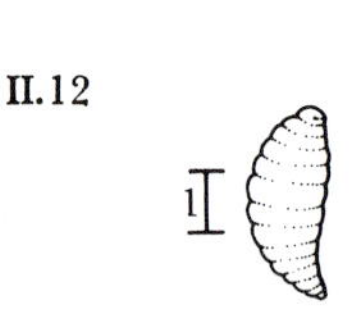

13 Larva active, with long hairs on each segment (II.9). TORYMIDAE *Torymus chloromerus* (Walker)

– Larva less active, not so hairy (II.10) 14

II.13

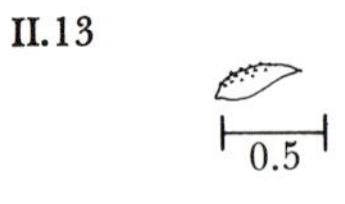

14 With large mandibles and several long hairs at head end (II.10); egg brown with a tail at both ends (II.11; egg-shell often obvious in gall-cell, tails may break off); common. EURYTOMIDAE *Eurytoma robusta* Mayr

– With small mandibles and ± hairless (II.12); egg white with papillae, rounded at both ends (II.13); common. PTEROMALIDAE *Pteromalus elevatus* (Walker)

15 External parasite of weevil *Apion* sp. PTEROMALIDAE *Trichomalus gynetelus* (Walker)

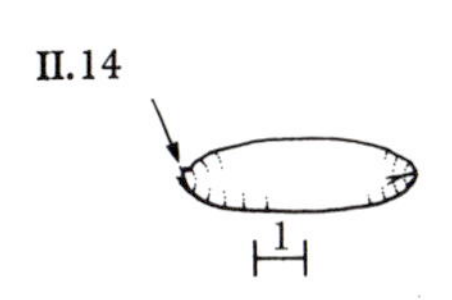

– Internal parasite of agromyzid *Melanagromyza aeneoventris*, often in puparium (II.14). PTEROMALIDAE *Syntomopus incisus* Thomson (Sometimes braconids (ICHNEUMONOIDEA) emerge from puparia of *M. aeneoventris.*)

– Parasite of caterpillar other CHALCIDOIDEA

16 Pupa inside puparium (segmented, II.4). DIPTERA: CYCLORRHAPHA 17

– Pupa free or inside silken cocoon; if in a brown cocoon, it is not segmented 20

17 Puparium brown (II.4), in gall-cell. TEPHRITIDAE *Urophora cardui* (L.)

– Puparium not in gall-cell 18

18 Puparium not as in II.14 other CYCLORRHAPHA

– Puparium with 2 spiracular processes at posterior end (II.14). AGROMYZIDAE 19

19 Puparium straw-coloured (II.14), embedded in pith of stem *Melanagromyza aeneoventris* (Fallèn)

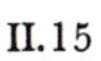

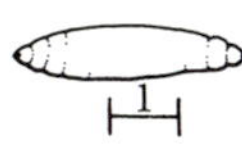

– Puparium reddish-brown (II.15) or otherwise, loose in stem other AGROMYZIDAE

20 Weevil pupa with elongate snout (II.3); not in cocoon or gall-cell. COLEOPTERA CURCULIONOIDEA (*Apion carduorum* Kirby and *A. onopordi* Kirby (APIONIDAE) are common in thistle stems.)

– Pupa without elongate snout; may be in cocoon or gall-cell 21

21 Parasite on or in remains of host; may be in puparium, gall-cell or silken or brown cocoon HYMENOPTERA

– Not parasitic; pupa free or in silken cocoon, usually with exit hole for adult cut in stem nearby. LEPIDOPTERA 22

22 Pupa pale brown, without cocoon; May–June. COCHYLIDAE *Aethes cnicana* (Westwood)

– Pupa in silken cocoon 23

23 Pupa in roots; May. COCHYLIDAE *Agapeta hamana* (L.)

– Pupa in stem 24

II.16

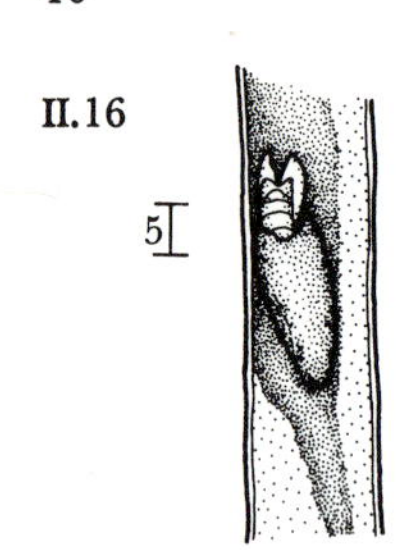

24 Pupal cocoon tough and papery, in upper part of stem (II.16); Apr.–May. TORTRICIDAE
Epiblema scutulana (D. & S.)

– Pupal cocoon a delicate net; Apr. PYRALIDAE
Myelois cribrella (Hübner)

III Insect larvae and pupae in leaves: hidden in leaf mines or silken webs, cocoons, folded leaves or spun-up shoots

III.1

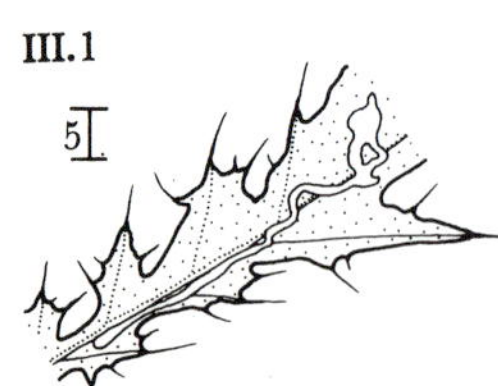

(Larvae and pupae are keyed together. Hymenopterous parasites are not included.)

1 Larva or pupa in leaf mine (III.1) 2

– Larva or pupa in silken web or cocoon or case, or folded leaf or spun-up shoot 13

III.2 III.3

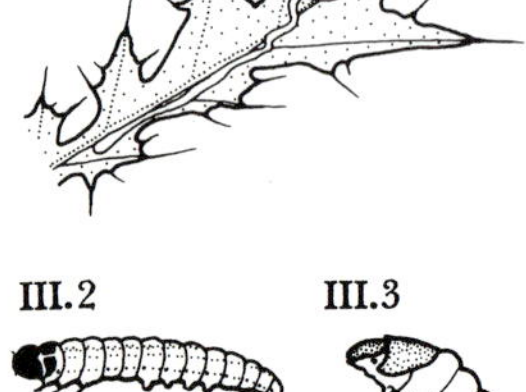

2 Larva with distinct head capsule and with legs (III.2, III.3); pupa free or in silken cocoon 3

– Larva with mouth-hooks, no distinct head capsule, without legs (III.4); pupa in segmented puparium (III.5). DIPTERA: CYCLORRHAPHA 9

III.4 III.5

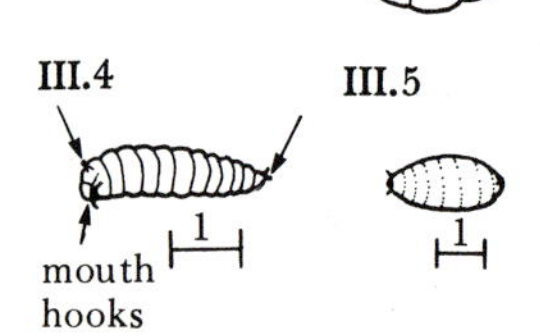

3 Larva with prolegs on abdomen (III.2); silk often present in mine. LEPIDOPTERA 4

– Larva without prolegs (III.3); no silk in mine unless pupa is present. COLEOPTERA: CHRYSOMELIDAE 7

III.6

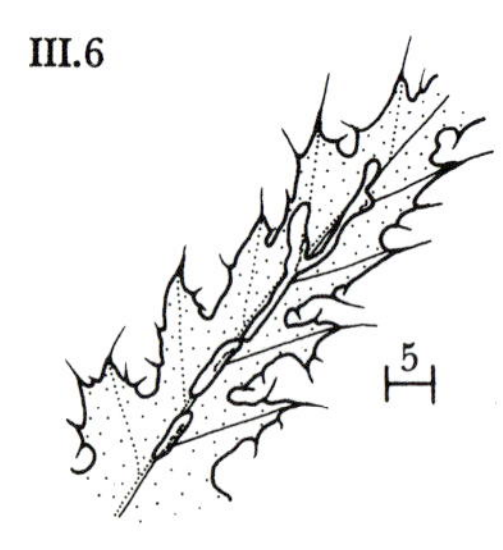

4 Linear mine (III.6) mainly in midrib of leaf 5

– Blotch mine (III.7) mainly in leaf blade, often associated with spun-up leaf or shoot 6

5 Mine without frass; larva pale pinkish green with darker longitudinal lines, head and prothoracic plate black; May–July. OECOPHORIDAE
Agonopterix carduella (Hübner)

– Mine with frass; larva grey-green, head and prothoracic plate brown; July and Sept. GELECHIIDAE
Scrobipalpa acuminatella (Sircom)

III.7

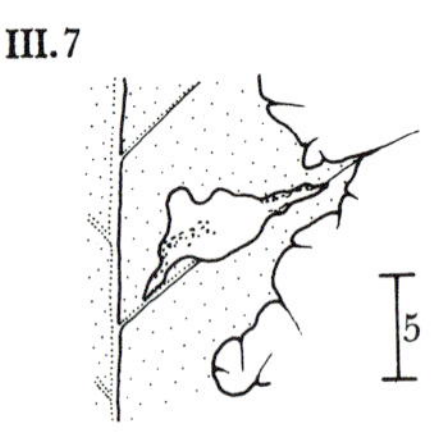

6 Larva Sept.–June; mining in first instar, older larva and pupa in spun-up shoots. TORTRICIDAE *Cnephasia* sp.

– Larva Apr.–Aug.; mining in first instar, older larva usually in web beneath leaf, pupa on ground.
OECOPHORIDAE other *Agonopterix* sp.

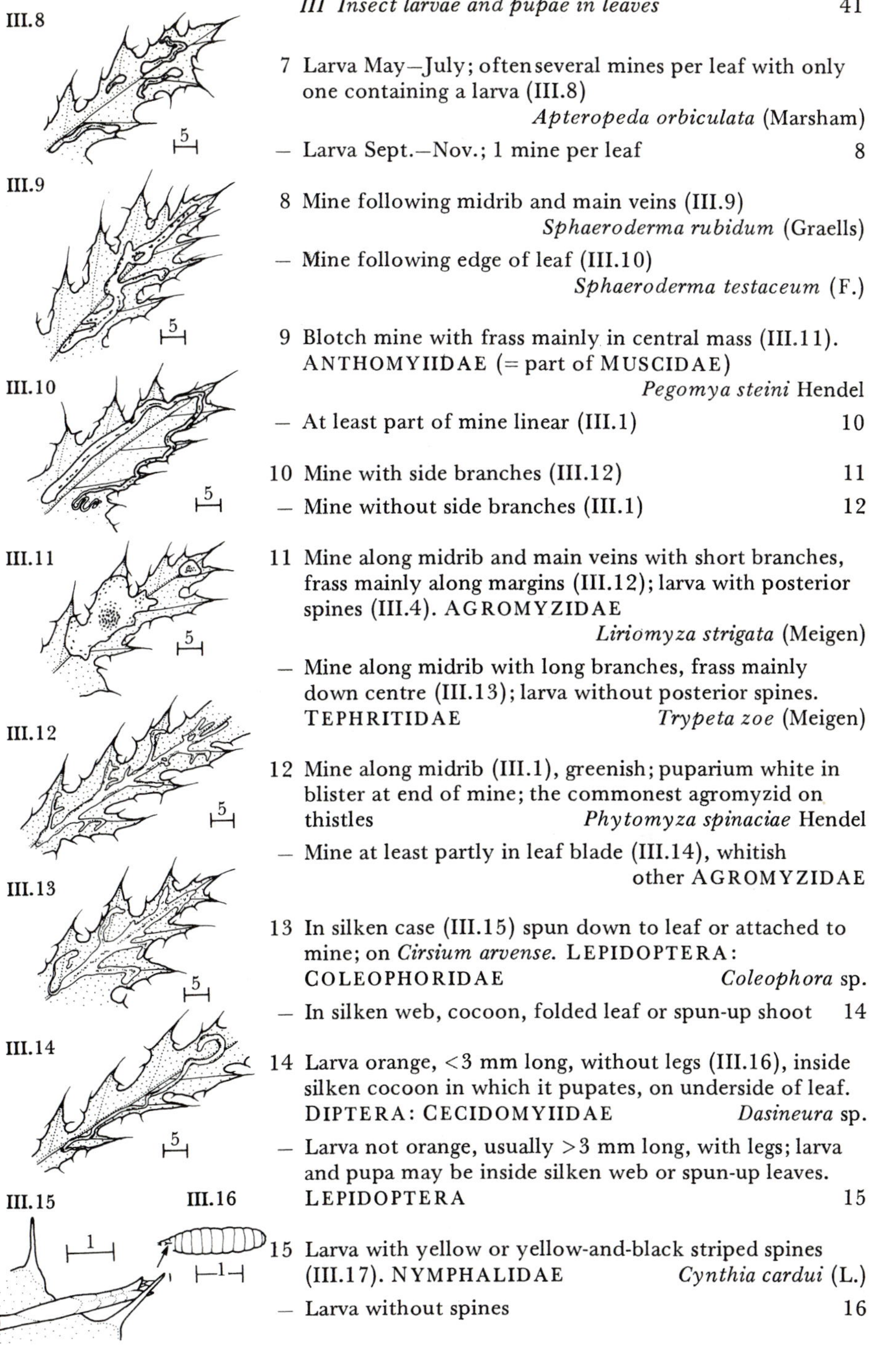

7 Larva May–July; often several mines per leaf with only one containing a larva (III.8)
Apteropeda orbiculata (Marsham)

– Larva Sept.–Nov.; 1 mine per leaf 8

8 Mine following midrib and main veins (III.9)
Sphaeroderma rubidum (Graells)

– Mine following edge of leaf (III.10)
Sphaeroderma testaceum (F.)

9 Blotch mine with frass mainly in central mass (III.11). ANTHOMYIIDAE (= part of MUSCIDAE)
Pegomya steini Hendel

– At least part of mine linear (III.1) 10

10 Mine with side branches (III.12) 11

– Mine without side branches (III.1) 12

11 Mine along midrib and main veins with short branches, frass mainly along margins (III.12); larva with posterior spines (III.4). AGROMYZIDAE
Liriomyza strigata (Meigen)

– Mine along midrib with long branches, frass mainly down centre (III.13); larva without posterior spines. TEPHRITIDAE *Trypeta zoe* (Meigen)

12 Mine along midrib (III.1), greenish; puparium white in blister at end of mine; the commonest agromyzid on thistles *Phytomyza spinaciae* Hendel

– Mine at least partly in leaf blade (III.14), whitish
other AGROMYZIDAE

13 In silken case (III.15) spun down to leaf or attached to mine; on *Cirsium arvense.* LEPIDOPTERA: COLEOPHORIDAE *Coleophora* sp.

– In silken web, cocoon, folded leaf or spun-up shoot 14

14 Larva orange, <3 mm long, without legs (III.16), inside silken cocoon in which it pupates, on underside of leaf. DIPTERA: CECIDOMYIIDAE *Dasineura* sp.

– Larva not orange, usually >3 mm long, with legs; larva and pupa may be inside silken web or spun-up leaves. LEPIDOPTERA 15

15 Larva with yellow or yellow-and-black striped spines (III.17). NYMPHALIDAE *Cynthia cardui* (L.)

– Larva without spines 16

III.17

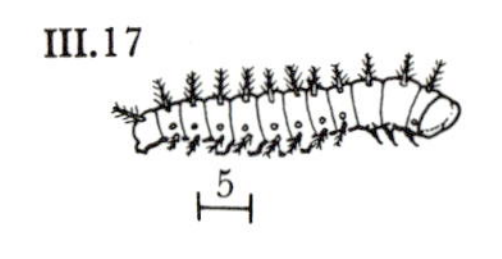

III.18

IV.1

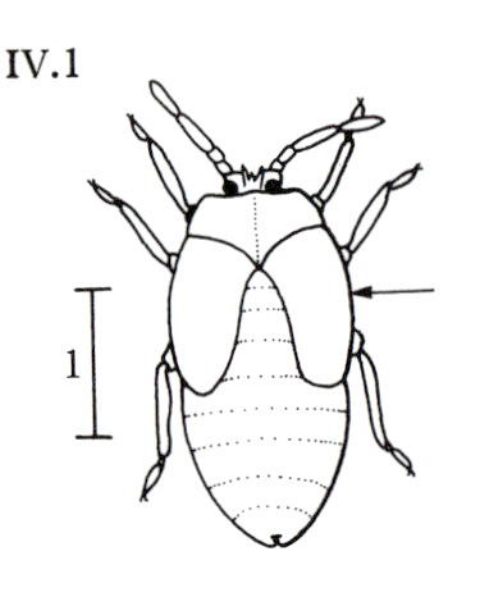

IV.2 IV.3

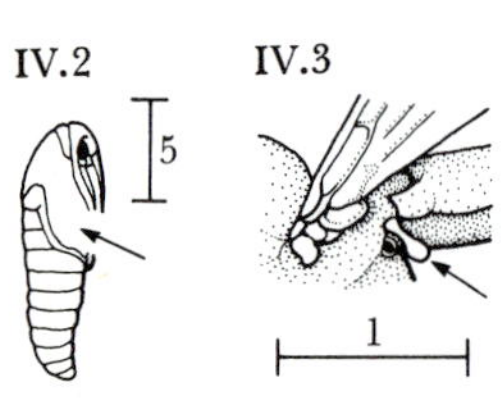

IV.4 IV.5

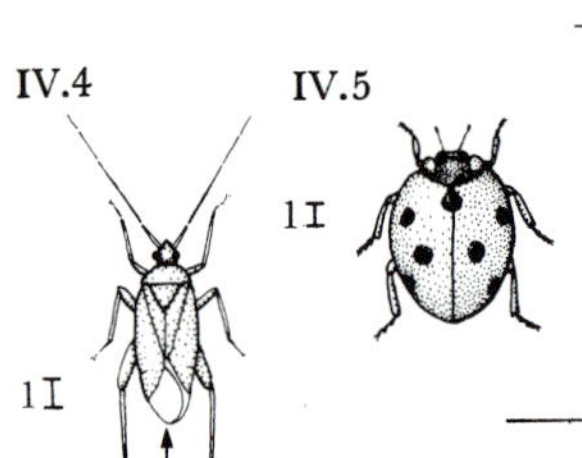

IV.6

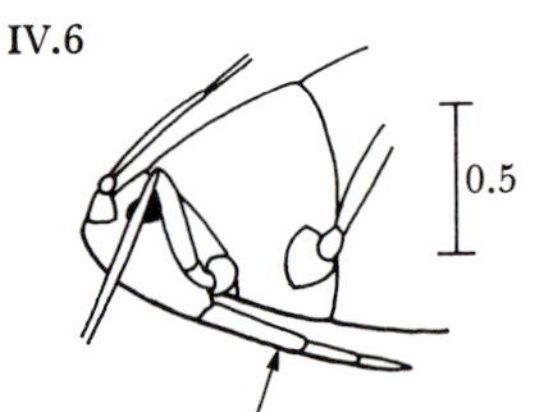

IV.7

16 Larva in folded or rolled leaf or in web on underside of leaf; spring–summer (fully fed in Aug.). OECOPHORIDAE — 17

– Larva in spun-up shoots, sometimes burrowing into head; summer–late spring (fully fed in June) — 18

17 Head and prothoracic plate yellowish brown, the latter with a black patch on either side; pinacula black (III.18) — *Agonopterix arenella* (D. & S.)

– Head and prothoracic plate usually dark; pinacula dark or same colour as abdomen — other *Agonopterix* sp.

18 Head olive brown, prothoracic and anal plates paler; abdomen bluish green with pinacula of same colour, shining; June and Aug.–Sept.; local in S. England, on *Cirsium arvense*. TORTRICIDAE — *Lobesia abscisana* (Doubleday)

– Head light to dark brown or black, prothoracic and anal plates brown or black; abdomen green or blue-grey with black pinacula; usually Sept.–June (if July–Aug. can be distinguished by pinacula) — 19

19 Head pale brown with dark spots; does not wriggle violently when disturbed. PYRALIDAE — *Sitochroa verticalis* (L.)

– Head pale to dark brown without spots; wriggles backwards violently when disturbed. TORTRICIDAE — *Cnephasia* sp.

(Caterpillars not fitting couplets 15–19 are 'other Lepidoptera': for further identification, see Buckler 1890–9; Meyrick, 1928; Beirne, 1954; Bradley *et al.*, 1973, 1979; Emmet, 1979.)

IV Insect adults, larvae and pupae on the outside of the plant, not protected by silk or folded leaves

A key to orders is given first: some of these are taken further.

1 With wings, sometimes first pair hardened concealing membranous second pair — 2

– With wing-buds (IV.1, IV.2) or without wings — 10

2 Forewings membranous, hindwings modified into club-shaped halteres (IV.3) — (flies) DIPTERA (key IV.3)

– 2 pairs of wings, forewings may be hardened — 3

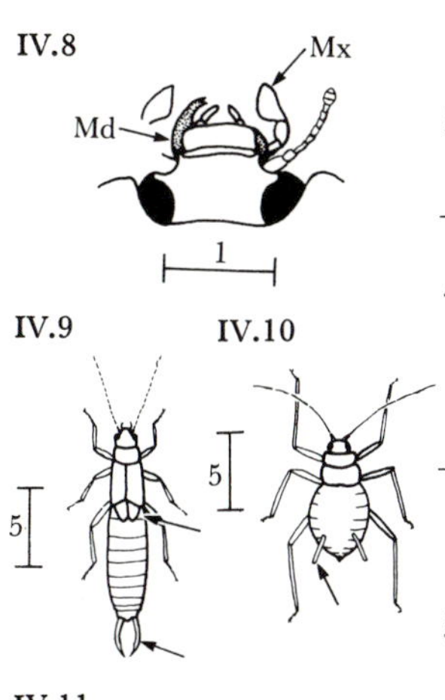

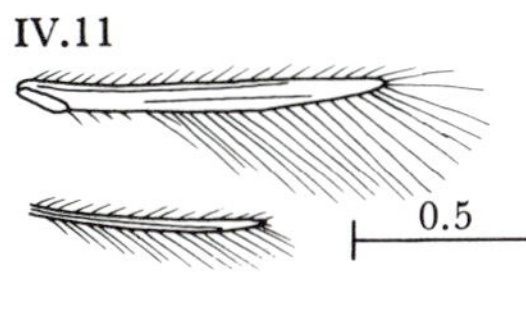

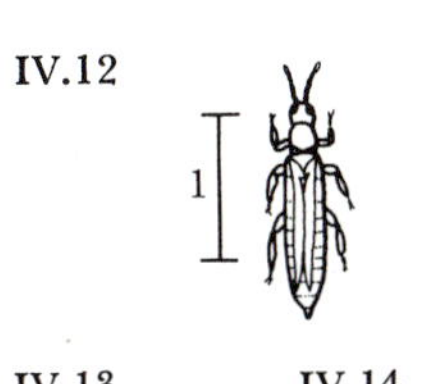

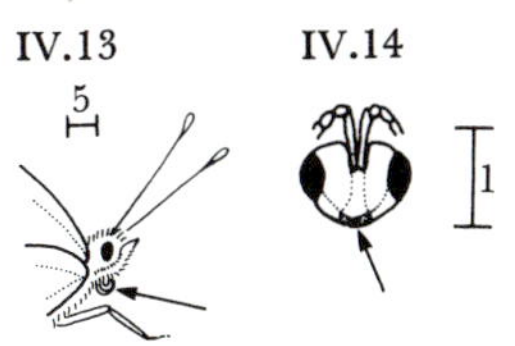

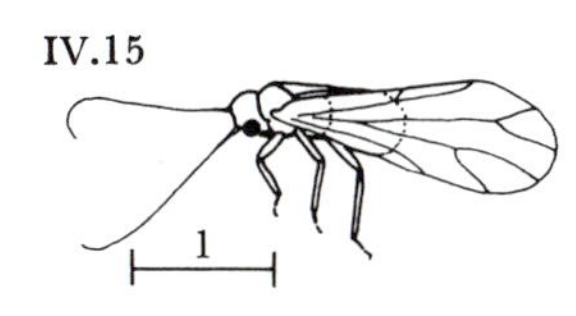

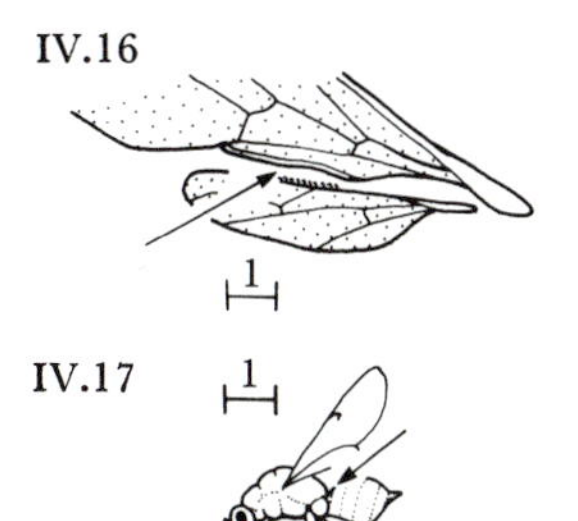

3 Forewings hardened covering membranous hindwings (IV.4, IV.5) — 4

– Both pairs of wings membranous — 6

4 Mouthparts prolonged into a sucking proboscis (rostrum, IV.6); forewings overlapping with membranous tips (IV.4) or held roof-like over body (IV.7)
(bugs) HEMIPTERA (key IV.1)

– Mouthparts with biting mandibles (IV.8: Md); forewings completely hardened and meeting down dorsal midline (IV.5) — 5

5 Forewings short, abdomen ending in forceps (IV.9)
(earwigs) DERMAPTERA

– Forewings (elytra) usually cover all or most of abdomen, no forceps at posterior end (IV.5)
(beetles) COLEOPTERA (key IV.4)

6 Mouthparts prolonged into a sucking proboscis (rostrum, IV.6); abdomen with 2 siphunculi (IV.10).
HEMIPTERA: HOMOPTERA
(aphids) APHIDOIDEA (key IV.1, couplet 10)

– Mouthparts not as in IV.6; no siphunculi — 7

7 Wings narrow fringed with hairs (IV.11), held flat over abdomen (IV.12) (thrips) THYSANOPTERA

– Wings not as in IV.11 — 8

8 Wings and body covered with flattened scales; mouthparts, if present, a coiled proboscis (IV.13)
(moths, butterflies) LEPIDOPTERA (key IV.2)

– Wings without flattened scales, though may be hairy; mouthparts with biting mandibles (IV.14) — 9

9 Wings held roof-like over abdomen, extending beyond its tip; antennae at least as long as body (IV.15); no waist between thorax and abdomen (psocids) PSOCOPTERA

– Wings held flat over abdomen, linked by hooks in a fold (IV.16); antennae shorter than body; abdomen usually constricted to a waist in front (IV.17; not in sawflies)
(ants, bees, wasps) HYMENOPTERA

(Identification of Hymenoptera adults is not attempted here: try Askew (1968), Chinery (1976), Ferrière & Kerrich (1968), Graham (1969) and Richards (1977). Some parasitoids can be identified as larvae on specific hosts (see keys I and II). If you intend to publish work on parasitoids, your identifications must be checked by an expert.)

IV.18

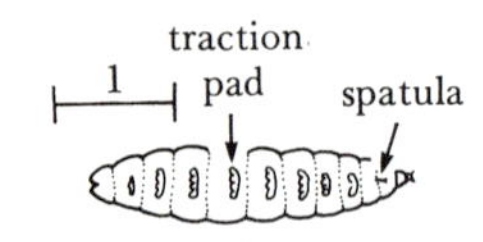

10 Inactive pupa with wing-buds and appendages 'glued' down to body (IV.2)
LEPIDOPTERA or COLEOPTERA
(Keep until adult emerges and identify using this key.)

— Active larva or nymph or adult ant, not as in IV.2 — 11

IV.19

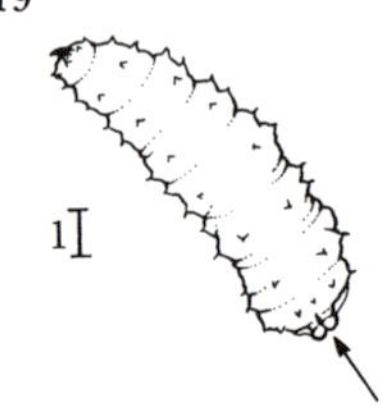

11 A maggot without legs; often in or near an aphid colony. DIPTERA — 12

— With thoracic legs and may have abdominal prolegs; not a maggot — 13

12 Red or orange, 3 mm or less long; with sternal spatula in fully grown larva and traction pads on ventral surface (IV.18). CECIDOMYIIDAE — *Aphidoletes* sp.

— Fleshy, green or cream-coloured often with dark gut, >3 mm long; 2 spiracles on projections at posterior end (IV.19) — (hover-flies) SYRPHIDAE

IV.20

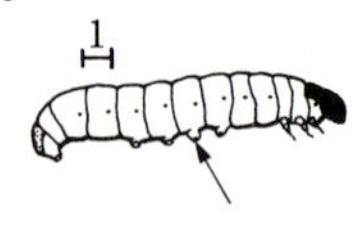

13 A caterpillar with prolegs on abdomen (IV.20)
(moths) LEPIDOPTERA (key IV.2)

— No prolegs on abdomen; not a caterpillar — 14

IV.21

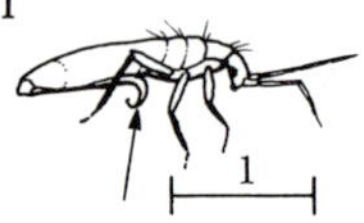

14 With forked springing organ attached to tip of abdomen and usually projecting forwards (IV.21)
(springtails) COLLEMBOLA

— Without a springing organ as in IV.21 — 15

IV.22

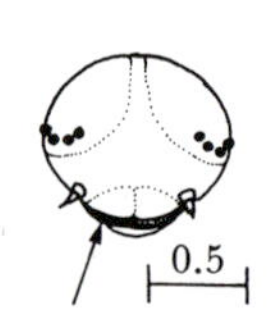

15 With biting mandibles (IV.22), mouthparts not prolonged into a proboscis — 16

— Mouthparts a sucking proboscis (rostrum, IV.6, IV.23); if scale-like, remove from plant to see rostrum — 20

16 Adult ant with abdomen narrow in front with outgrowths (IV.24). HYMENOPTERA
(ants) FORMICOIDEA

IV.23

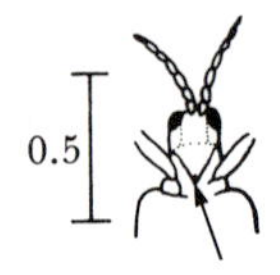

— Larva or nymph, abdomen not as in IV.24 — 17

17 Larva, without wing-buds; antennae shorter than body (IV.25) — 18

— Nymph, with wing-buds (IV.26, not on very young nymphs); antennae at least as long as head — 19

IV.24

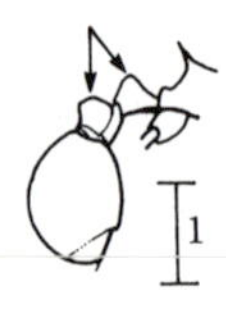

IV.25

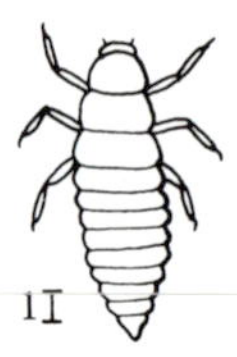

IV.26

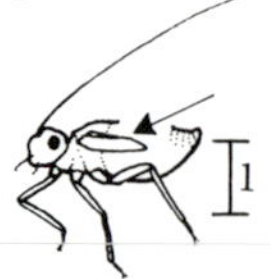

IV.27

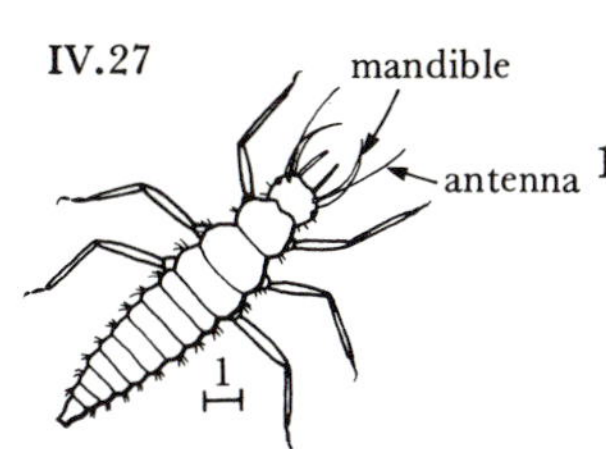

18 With large sickle-shaped mandibles extending in front of head, antennae longer than head; with tufts of hairs on each body segment (IV.27). NEUROPTERA (lacewings) CHRYSOPIDAE

– Mandibles not as in IV.27; antennae shorter than head; body without tufts of hairs but may be spiny (IV.28) (beetles) COLEOPTERA (key IV.4)

IV.28

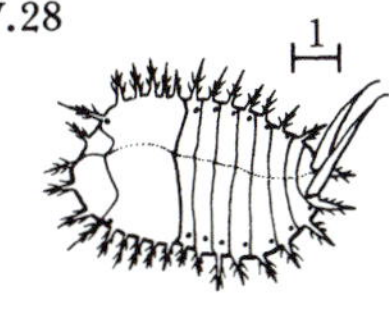
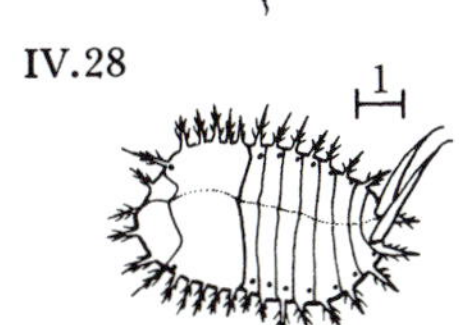

19 Body elongate, abdomen ending in forceps (IV.29); antennae shorter than body (earwigs) DERMAPTERA

– Body squat, abdomen without forceps; antennae longer than body (IV.26) (psocids) PSOCOPTERA

IV.29

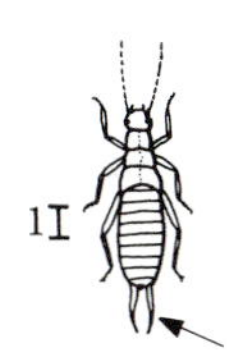

20 Rostrum short, head of characteristic shape (IV.23) (thrips) THYSANOPTERA

– Rostrum longer (IV.6); head not as in IV.23 (bugs) HEMIPTERA (key IV.1)

IV.1 HEMIPTERA ADULTS AND NYMPHS

IV.1.1

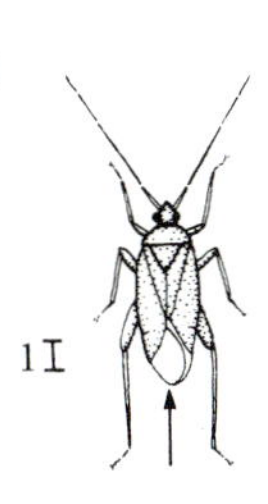

1 With wings, forewings may be hardened 2

– Without wings 21

2 Forewings partly hardened with membranous tips, overlapping and held flat over body (IV.1.1). HETEROPTERA 3

– Forewings completely hardened and held roof-like over body (IV.1.2), or membranous (IV.1.3). HOMOPTERA 8

IV.1.2

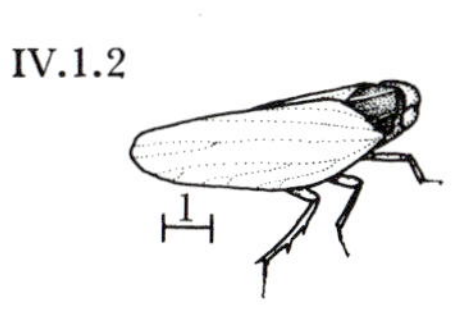

3 Rostrum held flat against body at rest (IV.1.4); mostly plant feeders 4

– Rostrum curved beneath body at rest (IV.1.5); carnivores 7

IV.1.3

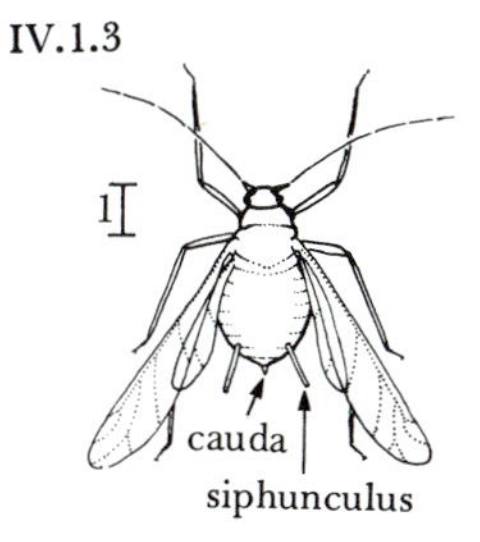

4 Prothorax and forewings with lacy net-like pattern, prothorax covering scutellum (triangle between bases of forewings, IV.1.6); body brown or grey with white powdery wax (may be rubbed off). TINGIDAE (lacebugs) 5

(PIESMIDAE are similar but prothorax does not cover scutellum.)

– Without net-like pattern as in IV.1.6; colours various 6

IV.1.4

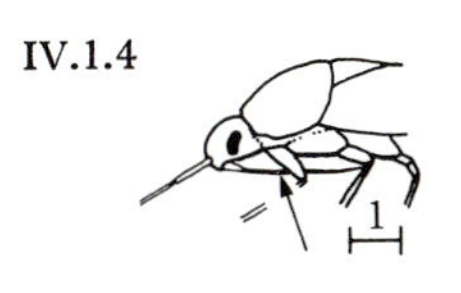

IV.1.5

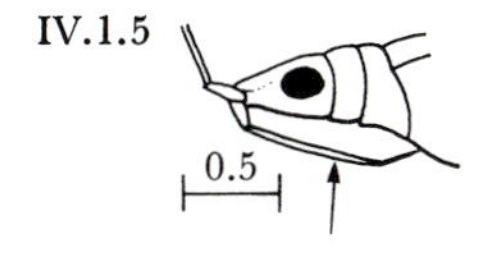

IV.1.6

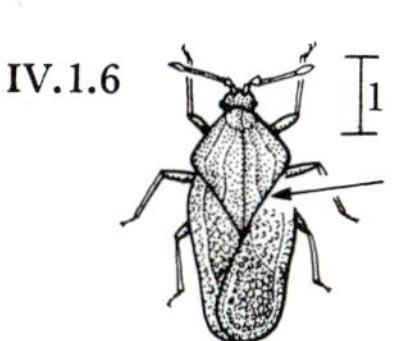

IV.1.7

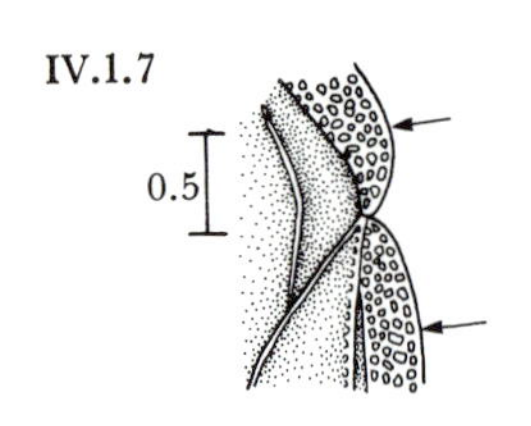

5 Usually on *Cirsium arvense*; lateral margins of prothorax and marginal area of forewing with 4 or 5 rows of meshes at widest point (IV.1.7); max. width >1.9 mm *Tingis ampliata* (Herrich-Schäffer)

– Usually on *C. vulgare*; lateral margins of prothorax and marginal area of forewing with 2 or 3 rows of meshes (IV.1.8); max. width <1.8 mm *Tingis cardui* (L.)

(*Tingis angustata* (Herrich-Schäffer) is rare on thistles in S. England; lateral margins of prothorax have 1 or 2 rows of meshes and it is narrower than *T. cardui.*)

IV.1.8

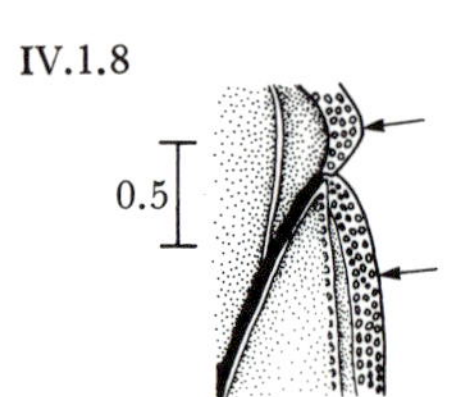

6 Tarsi 3-segmented; forewing with cuneus (IV.1.9); without ocelli (IV.1.10); >2.5 mm long, mainly green or brown MIRIDAE

– Not with this combination of characters other HETEROPTERA

IV.1.9 IV.1.10

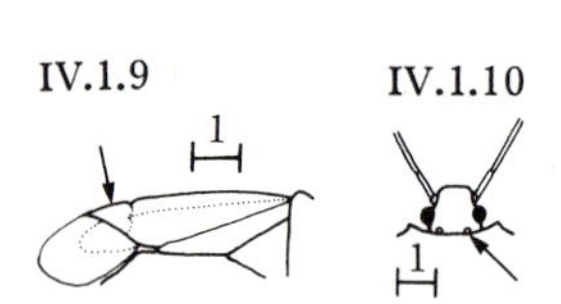

7 Rostrum with 3 segments; length <5 mm, as in pl. 4.7. CIMICIDAE Anthocorinae

– Rostrum with 3 or 4 segments; length usually >5 mm, not as in pl. 4.7 other HETEROPTERA

IV.1.11 IV.1.12

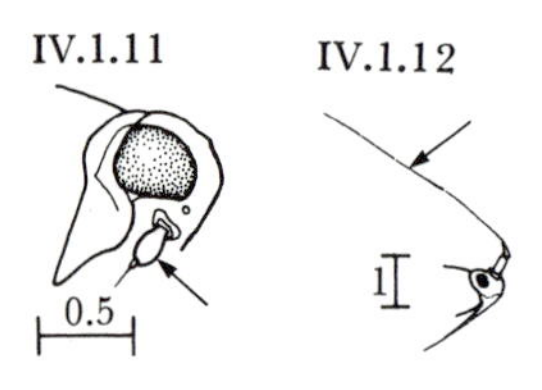

8 Antennae short with bristle at tip (IV.1.11); tarsi 3-segmented; forewings usually hardened (IV.1.2) (plant hoppers) AUCHENORRHYNCHA

– Antennae longer without bristle at tip (IV.1.12); tarsi with 1 or 2 segments; forewings usually membranous (IV.1.3). STERNORRHYNCHA 9

IV.1.13

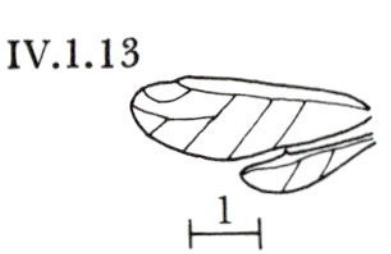

9 Tarsi with 2 segments of about equal length; venation not as in IV.1.13; no siphunculi other STERNORRHYNCHA

– Tarsi with 1, or 2 unequal, segments (IV.1.14); venation as in IV.1.13; usually 2 siphunculi on abdomen (IV.1.3), may be reduced to rings (IV.1.15). APHIDOIDEA 10

(Couplets 10–22 deal with aphids which feed on the thistles; casual visitors may be identified in Blackman, 1974.)

IV.1.14

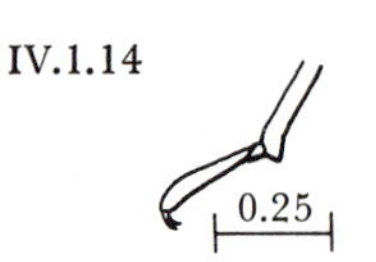

10 Hind tarsus $>\frac{1}{2}$ as long as hind tibia (IV.1.16), much longer than other tarsi; winged adult rare, greenish, wingless adult and nymphs more common, white; on roots, always attended by ants. LACHNIDAE 11

– All tarsi similar, $<\frac{1}{2}$ as long as hind tibia; greenish yellow, brown or black aphid on stem or leaves 12

IV.1.15

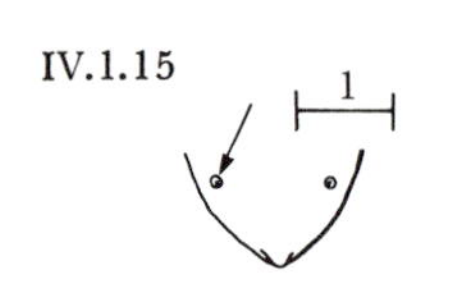

11 Siphunculi absent *Trama troglodytes* von Heyden

– Siphunculi present as a pair of dusky rings (IV.1.15) *Protrama radicis* (Kaltenbach)

IV.1.16

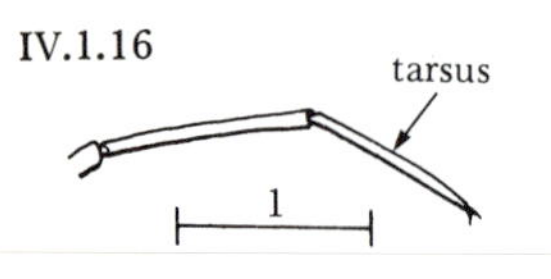

12 Antennae at least as long as body — 13
– Antennae shorter than body — 19

IV.1.17

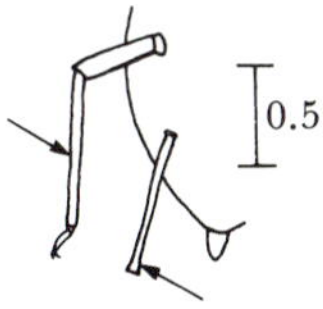

13 Siphunculi long, >½ length of hind tibia (IV.1.17); delicate inconspicuous aphid living under leaves or on 'wings' of stem. APHIDIDAE — 14
– Siphunculi <½ length of hind tibia — 15

14 Siphunculi yellow-white — *Capitophorus carduinus* (Walker)

IV.1.18

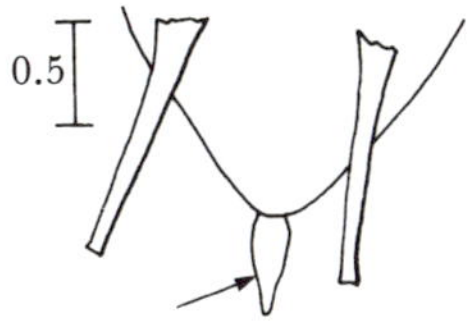

– Siphunculi with dusky tips — *Capitophorus eleagni* (Del Guercio)

15 Siphunculi dark brown; aphid dark red to bronzy black, clustering up stem and dropping off readily when disturbed. APHIDIDAE — 16
– Siphunculi pale; aphid yellowish or pinkish green, single on leaves or stem — 17

IV.1.19

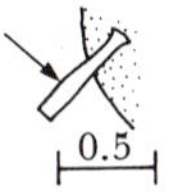

16 Cauda yellow (IV.1.18) — *Dactynotus cirsii* (L.)
– Cauda black — *Dactynotus aeneus* Hille Ris Lambers

IV.1.20

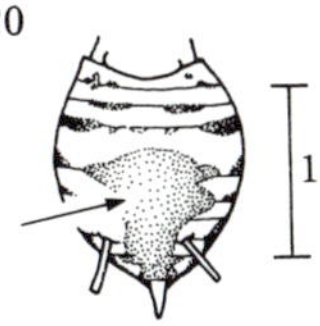

17 Siphunculi swollen (IV.1.19); winged adult with black patch on abdomen (IV.1.20) — *Myzus persicae* (Sulzer)
– Siphunculi not as in IV.1.19; no black patch on abdomen — 18

18 Siphunculi cylindrical ending in a flange (IV.1.21); often a dark green patch at base of each siphunculus. APHIDIDAE — *Aulacorthum solani* (Kaltenbach)

IV.1.21

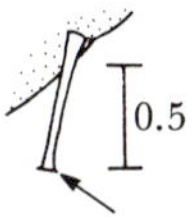

– Siphunculi not as in IV.1.21, without dark patch at base — other APHIDOIDEA

19 Not black or brown; living singly on stem or leaves — other APHIDOIDEA

IV.1.22

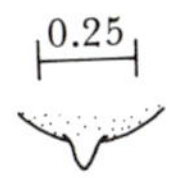

– Black or brown; living in a colony at top or base of stem; often attended by ants. APHIDIDAE — 20

20 Adult shiny black (nymphs green); cauda short, rounded, only slightly longer than wide (IV.1.22) — *Brachycaudus* sp.

IV.1.23

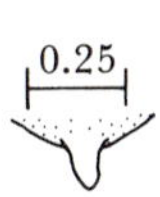

– Adult dull black or brownish (nymphs similar); cauda longer than wide (IV.1.23) — *Aphis* sp.

21 Usually with siphunculi (IV.1.3, IV.1.15); if none, hind tarsus >½ length of hind tibia (IV.1.16). HOMOPTERA: APHIDOIDEA — 22
– Without siphunculi — 23

IV.1.24

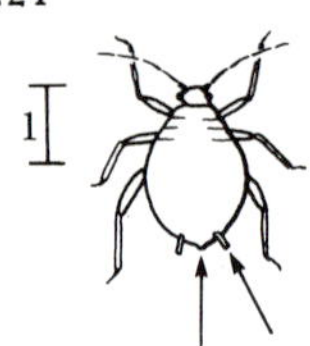

22 Siphunculi and cauda small, not fully formed (IV.1.24, pl. 4.2) immature APHIDOIDEA

– Siphunculi and cauda fully formed (IV.1.3); adult aptera (wingless aphid) 10
(Key as for alate (winged) aphids.)

23 Rostrum held flat against body at rest (IV.1.4); mostly plant feeders 24

– Rostrum curved beneath body (IV.1.5); carnivores. HETEROPTERA 27

24 Antennae short, bristle-like (IV.1.11); rostrum appears to arise from back of head. HOMOPTERA (plant hoppers) AUCHENORRHYNCHA

– Antennae long, not bristle-like; rostrum arises from front of head (IV.1.4). HETEROPTERA 25

25 Active, occurring singly, mainly green or brown; without warts; tarsi often with 3 segments other HETEROPTERA

– Inactive, occurring singly or in colonies, olive green; with pale warts (IV.1.25); tarsi with 2 segments. TINGIDAE (lacebugs) 26

IV.1.25

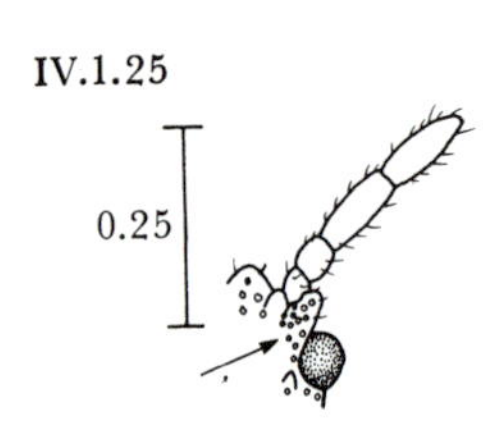

26 On *Cirsium arvense*, usually on vegetative bud or leaf *Tingis ampliata* (Herrich-Schäffer)

– On *Cirsium vulgare*, usually on bracts of flowering head or just below *Tingis cardui* (L.)

IV.2.1

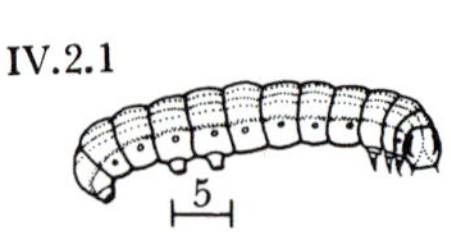

27 Rostrum with 3 segments; length <5 mm, as in pl. 4.8. CIMICIDAE Anthocorinae

– Rostrum with 3 or 4 segments; length usually >5 mm, not as in pl. 4.8 other HETEROPTERA

IV.2.2

IV.2.3 IV.2.4

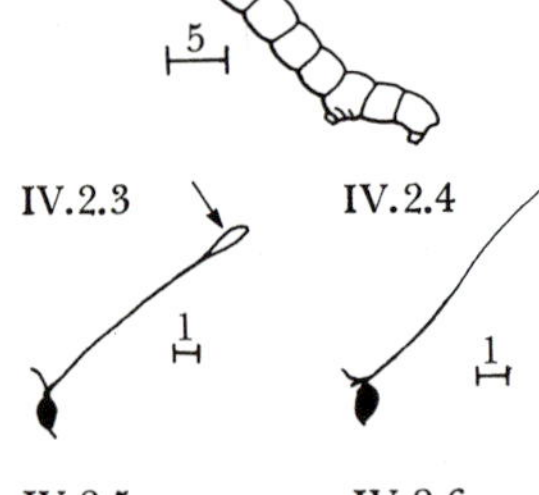

IV.2 LEPIDOPTERA ADULTS

Most larvae of Lepidoptera which feed on thistles are protected inside the heads, stems or leaf mines or in spun-up leaves and can be identified using keys I, II or III. Occasionally, larvae of NOCTUIDAE (IV.2.1), GEOMETRIDAE (IV.2.2) or other families feed exposed on the leaves; to identify further, see **Buckler, 1899**; Meyrick, 1928; Beirne, 1954; South, 1961; Bradley *et al.*, 1973, 1979; Jacobs *et al.*, 1978; **Carter, 1979**; Emmet, 1979.

Species belonging to five superfamilies feed and breed in thistles; their adults can be separated using table IV.2.1. For other adults feeding or resting on flowers, try Higgins & Riley (1970), Chinery (1976) and the references above.

IV.2.5 IV.2.6

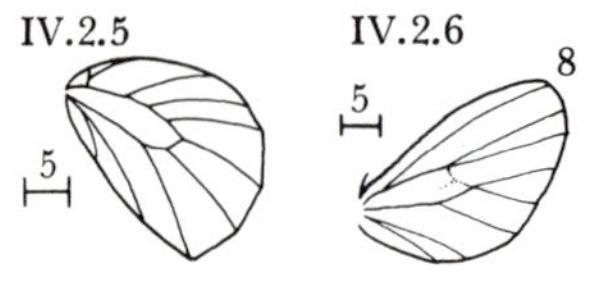

IV.2.7

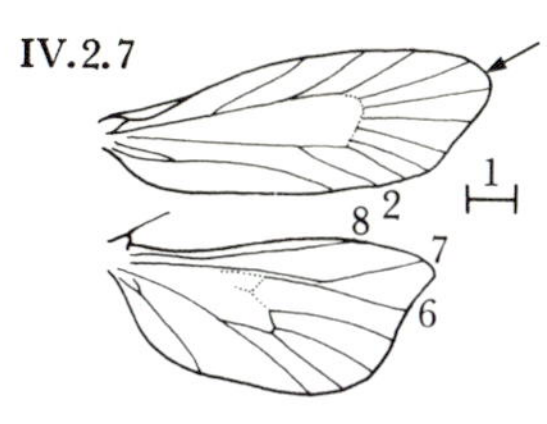

IV.2.8

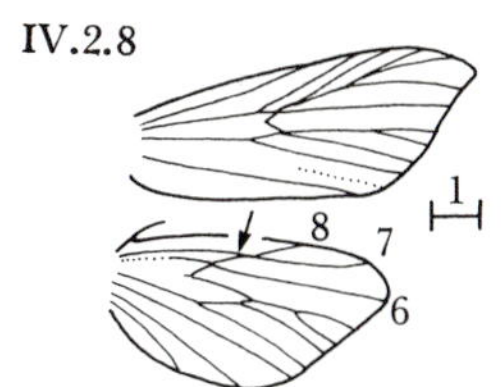

IV.2.9

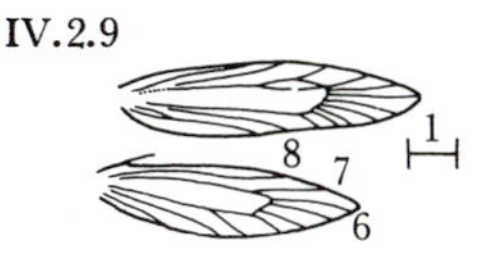

IV.2.10

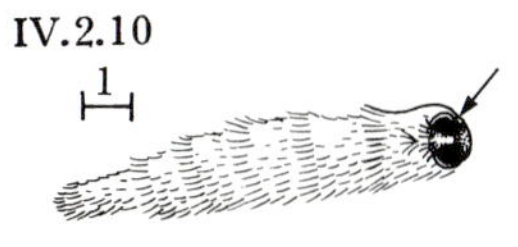

IV.2.11

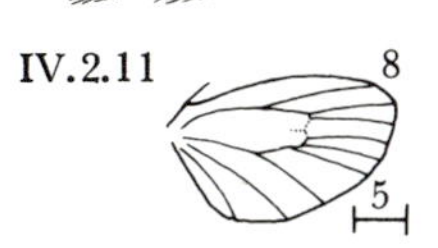

IV.2.1 Papilionoidea

Many butterflies rest on thistles or visit the flowers to feed on nectar. To identify, see Higgins & Riley (1970). The Painted lady *Cynthia cardui* (L.) (NYMPHALIDAE, fig. 38, p. 23) feeds and breeds on thistles.

IV.2.2 Noctuoidea

Many noctuoid moths may be found resting and occasionally feeding on thistles. NOCTUIDAE can be recognised by the venation of the hindwing (IV.2.6, vein 8 fuses with the central cell proximally; compare the hindwing of ARCTIIDAE, IV.2.11). Common species are the Silver Y *Autographa gamma* (L.) and the Sword-grass *Xylena exsoleta* (L.) (see South, 1961).

IV.2.3 Tortricoidea

Specimens not fitting this key are 'other Tortricoidea'; for further identification see Bradley *et al.* (1973, 1979).

1 Forewing with vein 2 arising beyond ¾ along central cell, vein 1C absent (IV.2.7). COCHYLIDAE 2

– Forewing with vein 2 arising before ¾ along central cell, vein 1C present at margin (IV.2.12). TORTRICIDAE 4

Table IV.2.1. *To identify adults belonging to five superfamilies of Lepidoptera*

	Papilionoidea	Noctuoidea	Tortricoidea	Pyraloidea	Tineoidea
Antennae:	Clubbed (IV.2.3)	Tapering	Tapering	Tapering (IV.2.4)	Tapering
Wing fringe:	Short	Short	Long	Long	Long
Ratio width hindwing to length fringe:	>8	>8	<5	<5	<5
Venation:[a]	IV.2.5 hindwing	IV.2.6 hindwing	IV.2.7 forewing ± square at outer edge, veins 6 & 7 of hindwing close together or stalked	IV.2.8 veins 7 & 8 fused beyond cell in hindwing	IV.2.9 veins 7 & 8 of hindwing usually separate
Usual attitude of wings at rest:	Vertically above body	Flat over body	Roof-wise over body	Flat over body	Flat over body
Hearing organs:	Absent	On metathorax	Absent	On abdomen (IV.2.10) (may need to remove abdomen)	Absent
Proceed to:	**key IV.2.1**	**key IV.2.2**	**key IV.2.3**	**key IV.2.4**	**key IV.2.5**

[a]To see veins without damaging scales, moisten wing on underside with a drop of alcohol or ether.

IV.2.12

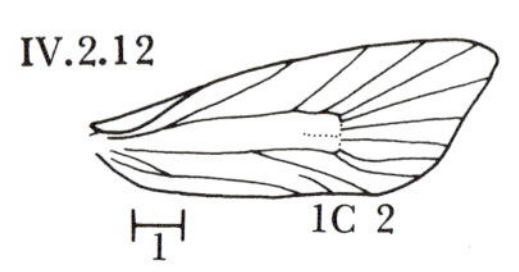

IV.2.13

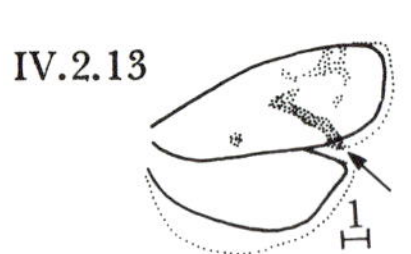

IV.2.14

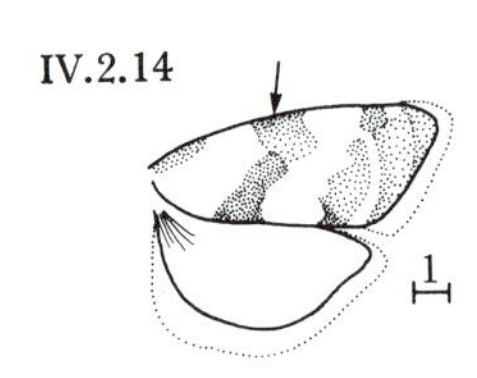

IV.2.15

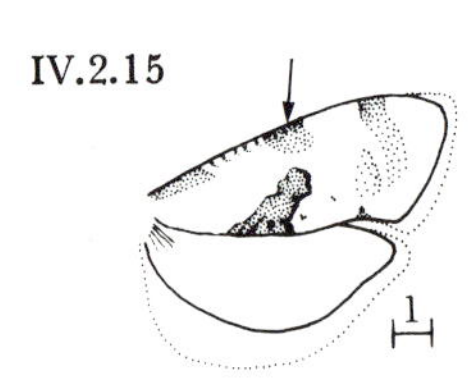

IV.2.16

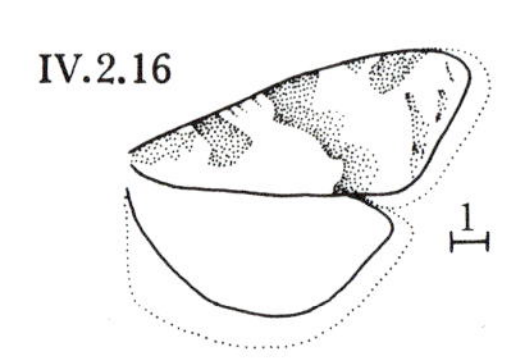

IV.2.17 IV.2.18

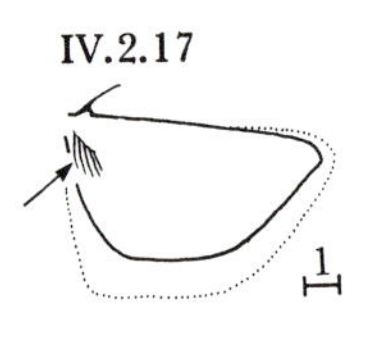

IV.2.19

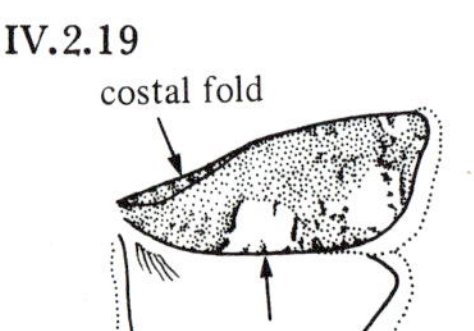

IV.2.20

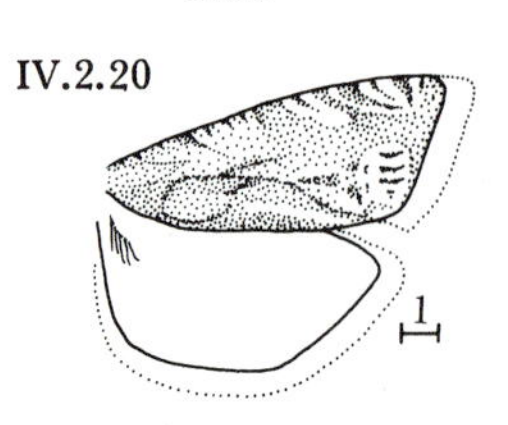

2 Ground colour of forewing uniformly yellow with a brown streak from centre to outer posterior corner extending into fringe (IV.2.13) *Agapeta hamana* (L.)

– Ground colour of forewing white with yellow or grey, with ± interrupted brown bands (IV.2.14) 3

3 Middle band of forewing almost complete, slightly oblique (IV.2.14); usually on thistles (pl. 3.4) *Aethes cnicana* (Westwood)

– Middle band of forewing clearly interrupted above middle, distinctly oblique (IV.2.15); usually on burdocks *Aethes rubigana* (Treitschke)

4 Hindwing without pecten; forewing with ± distinct oblique grey-brown or grey-black bands (IV.2.16). Tortricinae *Cnephasia* sp.

– Hindwing with pecten on upper surface (IV.2.17). Olethreutinae 5

5 Wing span 10–13 mm; forewing white-grey with ± distinct brown bands (IV.2.18); ♂ forewing without costal fold; May and July–Aug., local in S. England *Lobesia abscisana* (Doubleday)

– Wing span 12–23 mm; forewing not banded; ♂ forewing with costal fold (IV.2.19); widespread and common 6

6 Forewing with distinct cream patch in middle of posterior half, extending to outer margin in ♂ (IV.2.19 ♂, pls. 3.5, 3.6); May–June *Epiblema scutulana* (D. & S.)

– Forewing with poorly defined markings, varying from pale with brown streaks to dark with pale streaks (IV.2.20; pl. 3.7); June–Aug. *Eucosma cana* (Haworth)

IV.2.4 Pyraloidea

Specimens not fitting this key are 'other Pyralidae'; for further identification see Beirne (1954).

1 Each wing divided into 2 or 3 lobes (IV.2.21) PTEROPHORIDAE

– Wings not divided. PYRALIDAE 2

2 Hindwing without pecten; wings yellow-cream, forewing with oblique narrow bands of yellow-orange, hindwing with dark brown bands fading towards posterior margin (IV.2.22). Pyraustinae *Sitochroa verticalis* (L.)

– Hindwing with pecten on upper surface (IV.2.17); wings white, grey or brownish without bands, forewing with brown or black spots. Phycitinae 3

IV.2.21

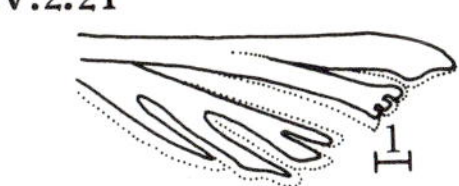

IV.2.22

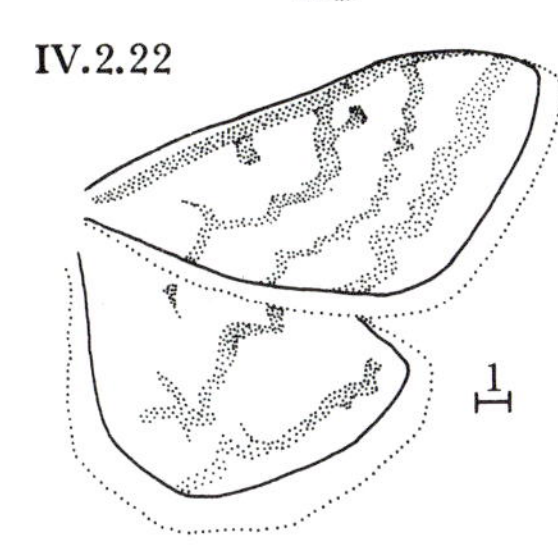

IV.2.23

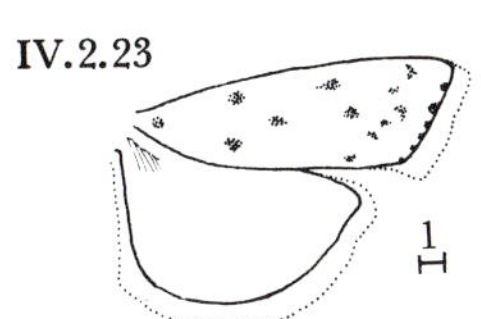

IV.2.24

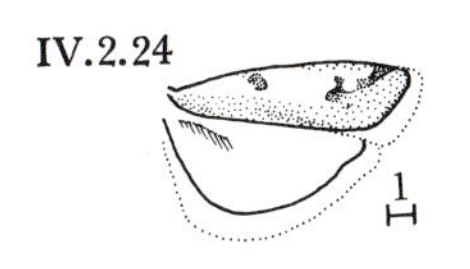

IV.2.25

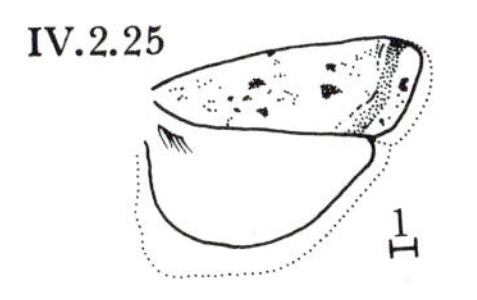

IV.2.26 IV.2.27

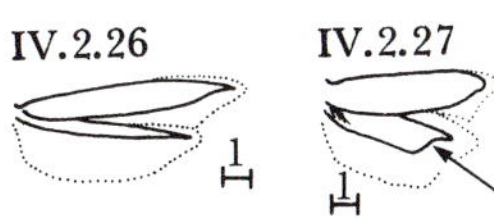

IV.2.28 IV.2.29

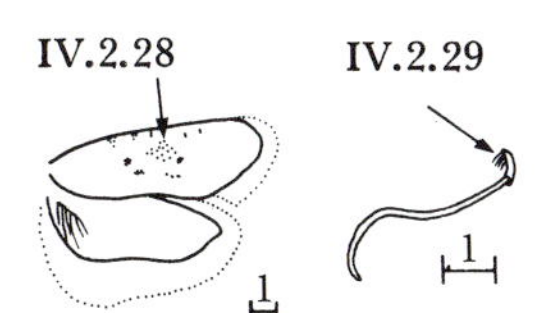

IV.3.1

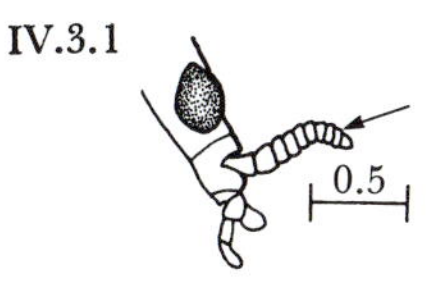

IV.3.2

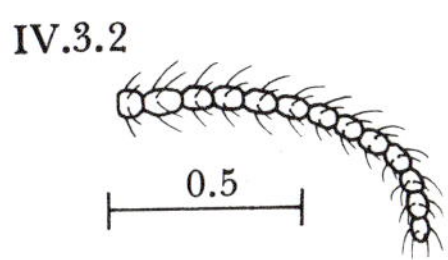

3 Wing span >25 mm; forewing white with black spots, hindwing white suffused with grey-brown, fringe white (IV.2.23, pl. 3.1) *Myelois cribrella* (Hübner)

– Wing span <25 mm; forewing cream suffused with pale brown with dark brown spots, hindwing pale brown, fringe cream (IV.2.24) 4

4 Wing pattern IV.2.24, colour of hindwing uniform except for dark brown edge (pl. 3.9) *Phycitodes binaevella* (Hübner)

– Wing pattern IV.2.25, colour of hindwing darkens towards tip *Homoeosoma nebulella* (D. & S.)

IV.2.5 Tineoidea

Specimens not fitting this key are 'other Tineoidea'; for further identification see Jacobs (1955), Chinery (1976), Emmet (1979).

1 Wings narrow and pointed, fringe of hindwing > twice width of hindwing (IV.2.26, pl. 3.2). COLEOPHORIDAE *Coleophora* sp.

– Wings less narrow, hindwing fringe < twice its width 2

2 Hindwing with curved posterior margin, its fringe > width of wing (IV.2.27, pl. 3.3); no tuft of hairs on antenna GELECHIIDAE

(The larva of *Scrobipalpa acuminatella* (Sircom) mines thistle leaves causing brown blotches.)

– Wings rounded, forewing with dark brown spots, fringe of hindwing less than width of wing (IV.2.28, pl. 3.8); antenna with tuft of hairs at base (IV.2.29). OECOPHORIDAE *Agonopterix* sp.

Five species of *Agonopterix* are known to breed in thistles. The pattern of spots and colour of the forewing differs slightly (see colour plate in Jacobs, 1955).

IV.3 DIPTERA ADULTS

(Larval syrphids and cecidomyiids are included in key IV; other fly larvae are not usually found on outside of thistle plants.)

1 Antennae with >5 segments (IV.3.1, IV.3.2) 2

– Antennae with <5 segments (IV.3.3) 3

2 Wing with 2–6 veins reaching margin (IV.3.4); no spurs at tip of tibia, usually first tarsal segment *c*. ⅕ length of second (IV.3.5); eyes often meet above antennae in a parallel-sided eye bridge (gall-midges) CECIDOMYIIDAE

– Without any or all of these characters other DIPTERA

IV.3.3

arista

0.5

IV.3.4

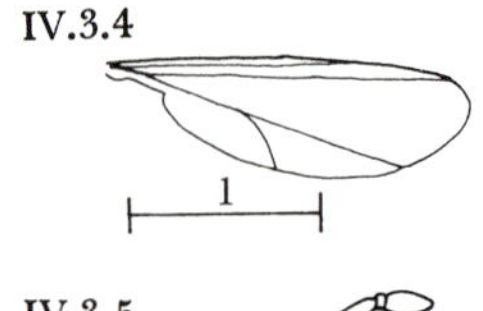

IV.3.5

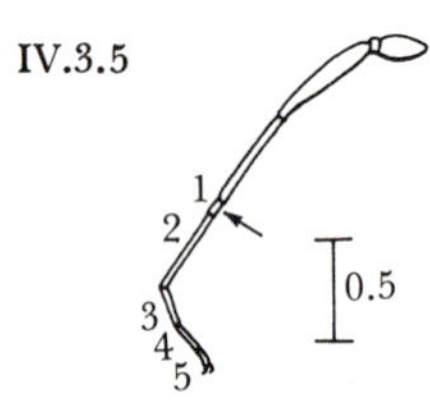

IV.3.6

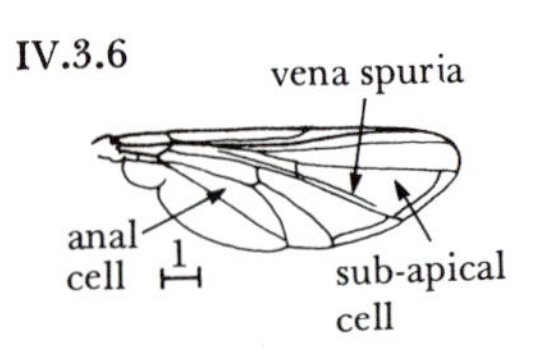

IV.3.7

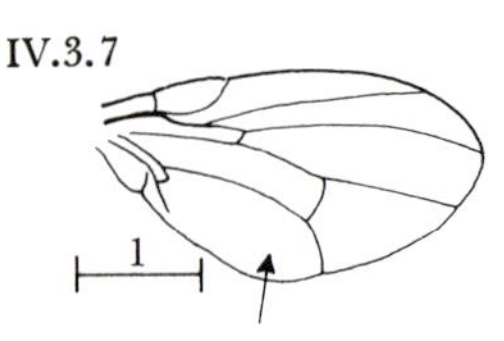

IV.3.8

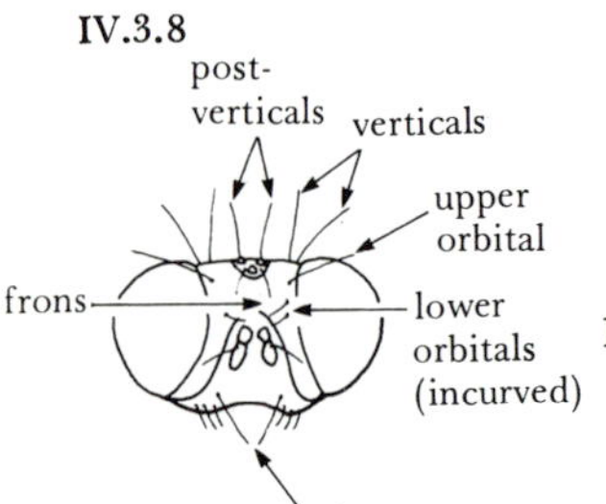

IV.3.9

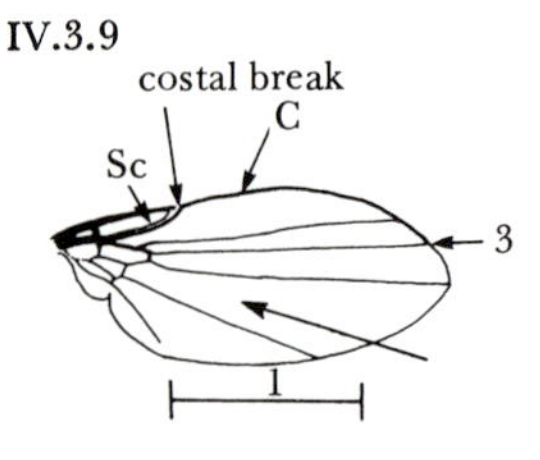

3 Anal cell of wing long, pointed, extending almost to margin (IV.3.6) — 4

– Anal cell short, extending at most halfway to margin, or absent (IV.3.7) — 5

4 Sub-apical cell of wing closed, vena spuria (not a true vein but a pigmented fold) present (IV.3.6); proboscis usually shorter than head — (hover-flies) SYRPHIDAE

– Without any or all of these characters — other DIPTERA

5 Antennae with 3 segments and dorsal arista (bristle, IV.3.3); tibia with no bristles longer than its width — 6

– Without either or both of these characters — other DIPTERA

6 Vibrissae present each side of mouth (IV.3.8) — 7

– Vibrissae absent — 8

7 Lower orbital bristles on head incurved (towards mid-line), post-vertical bristles divergent (pointing away from each other) (IV.3.8); sub-costal vein (Sc) of wing complete but fading before joining costa (C, IV.3.9), or Sc merges with vein 1 (radius, R) before joining costa (IV.3.10) — (leaf-miners) AGROMYZIDAE (key IV.3.1)

– Without any or all of these characters — other DIPTERA

8 Wing with 4 closed cells (first and second basal, anal, discal), anal vein extends at least ¼ way to wing margin (IV.3.11) — 9

– Wing with fewer veins — other DIPTERA

9 1–3 pairs of black, incurved lower orbital bristles (IV.3.8) — TEPHRITIDAE (key IV.3.2)

– Lower orbital bristles not incurved, or absent — 10

10 Wings patterned, vein 1 (R) without hairs on upper side (IV.3.12); post-vertical bristles present (IV.3.8); body not shining black — PALLOPTERIDAE (key IV.3.3)

– Without any or all of these characters — other DIPTERA

(To identify other Diptera, see Oldroyd, 1954; Colyer & Hammond, 1968; Chinery, 1976; **Unwin, 1981.**)

IV.3.1 Agromyzidae

Check family identification: size <4 mm, wings not patterned; post-vertical bristles divergent; third antennal segment rounded (IV.3.13) or produced to a sharp point; one costal break (IV.3.9).

IV.3.10

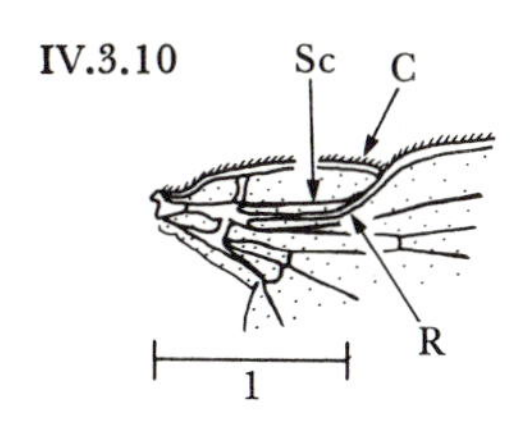

IV.3.11

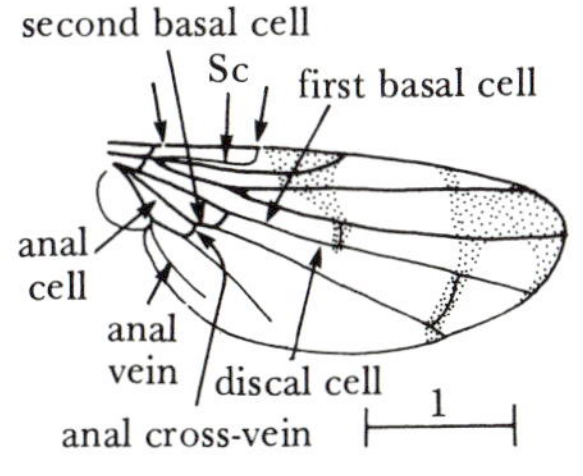

IV.3.12 IV.3.13

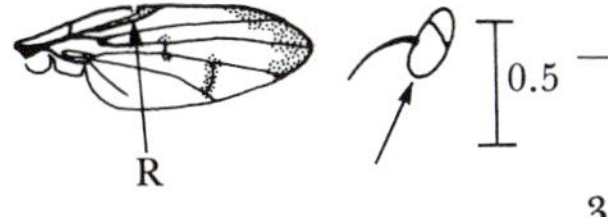

IV.3.14

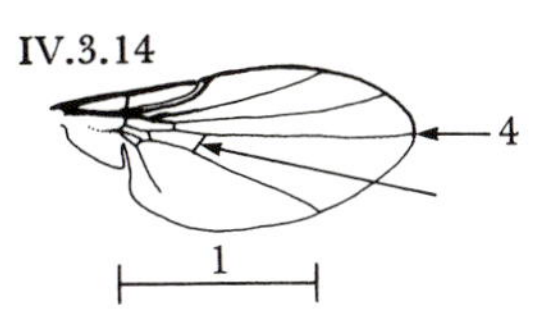

IV.3.15 IV.3.16

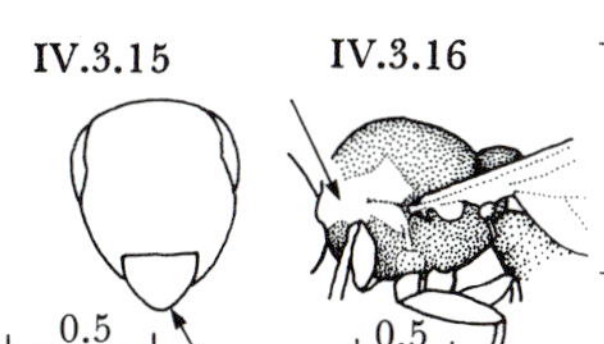

IV.3.17 IV.3.18

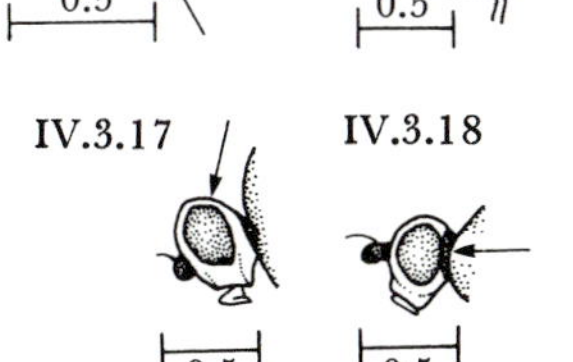

IV.3.19

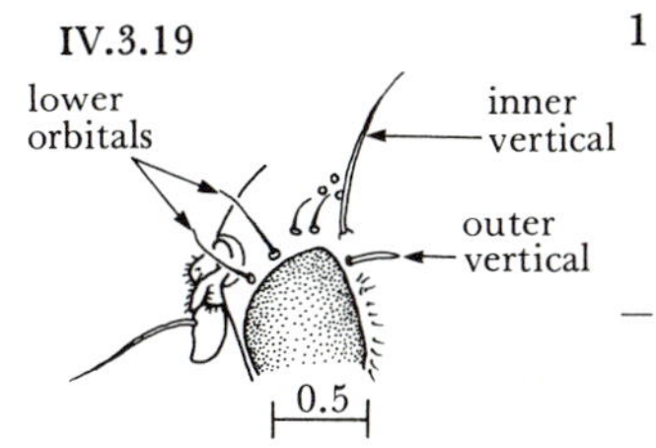

Agromyzids are a difficult group. This key should be reliable for specimens bred from the thistles but may not work for casual visitors; for these, see Spencer (1972, 1976). If you intend to publish work, all identifications should be checked by an expert.

1 Sub-costal vein (Sc) strong, joining radius (R) before the costa (C, IV.3.10); body shining greenish or coppery black; wing length 2.4–3.5 mm; larva a stem-borer, puparium pale straw-coloured. Agromyzinae *Melanagromyza aeneoventris* (Fallèn)

– Vein Sc weak, joining costa independently (IV.3.9); body grey, not shining; wing length often <2.5 mm; larva a leaf-miner, puparium brown. Phytomyzinae 2

2 Costa extending to apex of vein 4 (M_{1+2}), second cross-vein present (IV.3.14); scutellum yellow (IV.3.15) *Liriomyza* sp.

– Costa extending to apex of vein 3 (R_{4+5}), second cross-vein absent (IV.3.9); scutellum grey 3

3 Yellowish patch on side of thorax extending to anterior edge (IV.3.16) *Phytomyza cirsii* Hendel

– No yellow patch on side of thorax 4

4 Yellowish colour of head completely encircles eye (IV.3.17) *Phytomyza syngenesiae* (Hardy)

– Yellowish colour of head encircles eye except for dark posterior margin (IV.3.18) 5

5 Wing 3.3–3.8 mm long *Phytomyza continua* Hendel

– Wing <2.5 mm long; the commonest *Phytomyza* on thistles *Phytomyza spinaciae* Hendel

IV.3.2 Tephritidae

Check family identification: size >4 mm, usually with patterned wings; sub-costal vein with a right-angled bend (Sc, IV.3.11), often weak; two costal breaks (IV.3.11; difficult to see, often marked by stronger bristles).

1 Outer vertical bristle short and spine-like, similar to stiff bristles along posterior edge of eye, inner vertical bristle long and hair-like (IV.3.19); wing markings often a complicated pattern of dark patches and streaks (IV.3.20), not banded. Tephritinae 2

– Both vertical bristles long and hair-like (IV.3.21); wings with dark bands (IV.3.11) or spots (IV.3.22) or pattern absent 4

2 Wing markings faint (IV.3.23); 3 pairs lower orbital bristles (IV.3.21); larva in flower head gall
Acanthiophilus helianthi (Rossi)

– Wing markings distinct; 2 pairs lower orbital bristles (IV.3.19) 3

3 Wing pattern as in IV.3.24; rare, almost specific to *Cirsium arvense*; larva in flower head
Tephritis cometa (Loew)

– Wing pattern as in IV.3.20; common, on several *Cirsium* spp.; larva in flower head gall *Tephritis conura* (Loew)

4 Cross-vein closing anal cell of wing sharply bent (IV.3.25); 2 pairs upper orbital bristles (IV.3.21). Trypetinae 5

– Cross-vein closing anal cell flat or gently curved (IV.3.11); 1 pair upper orbital bristles (IV.3.26). Urophorinae 11

5 Wings clear, no dark bands or spots; larva in flower head
Terellia serratulae (L.)

– Wings patterned with bands or spots 6

6 Wing pattern as in IV.3.27, IV.3.28; the 2 upper orbital bristles lying ± parallel (IV.3.29); larva a leaf-miner
Trypeta zoe (Meigen)

– Wing pattern different; the 2 upper orbital bristles diverging (IV.3.30) 7

7 Vein 3 of wing with bristles along whole length (IV.3.22); frons with many fine hairs between the bristles (IV.3.8); wing pattern variable but never banded; larva in flower head *Xyphosia miliaria* (Schrank)

– Vein 3 bare or with bristles at junction with vein 2 only (IV.3.31); frons without fine hairs (occasionally with 1 or 2); wings often banded 8

8 Wings banded, though colour sometimes faint (IV.3.31); scutellum with a dark patch apically (IV.3.32) 9

– Wings with dark patches, never in complete bands (IV.3.33); scutellum entirely yellow 10

9 With a pair of bristles anterior to transverse suture on thorax, distinct black spots at bases of dorsal thoracic bristles (IV.3.32); larva in flower head gall
Chaetorellia jaceae (R.-D.)

– Without this pair of bristles, dorsal surface of thorax mainly dark including bases of bristles; larva in flower head *Chaetostomella onotrophes* (Loew)

IV.3.20

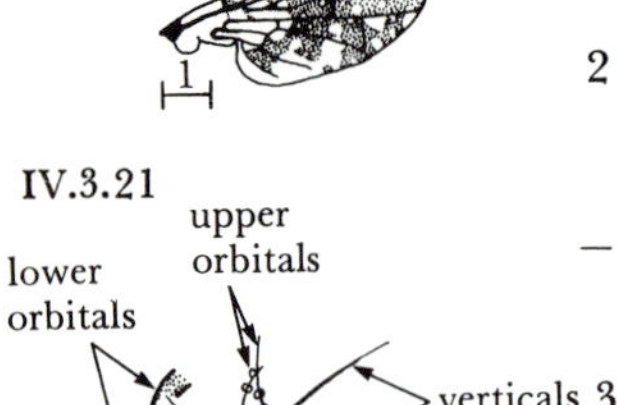

IV.3.21

IV.3.22

IV.3.23 IV.3.24

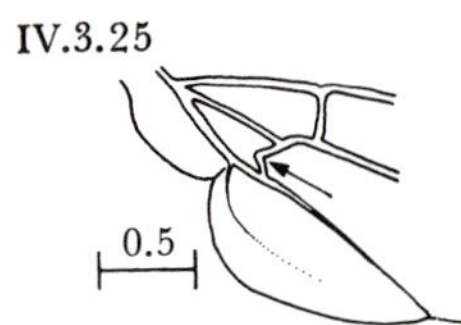

IV.3.25

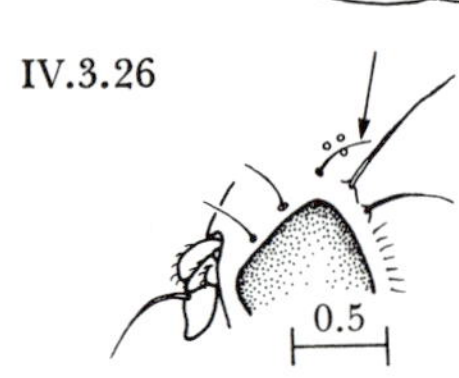

IV.3.26

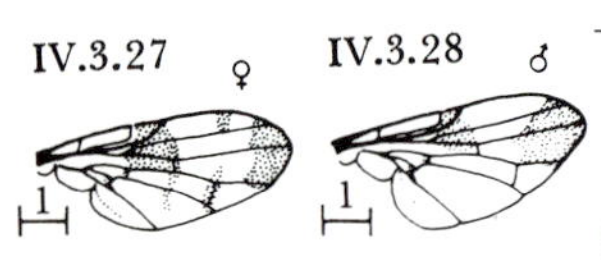

IV.3.27 ♀ IV.3.28 ♂

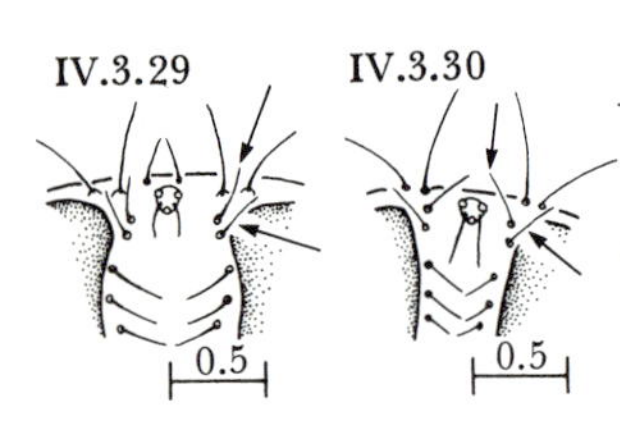

IV.3.29 IV.3.30

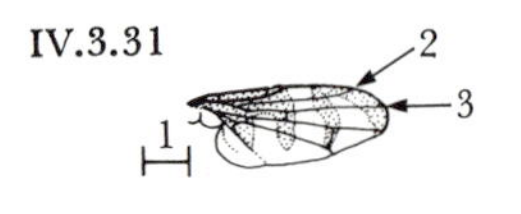

IV.3.31

IV.3.32

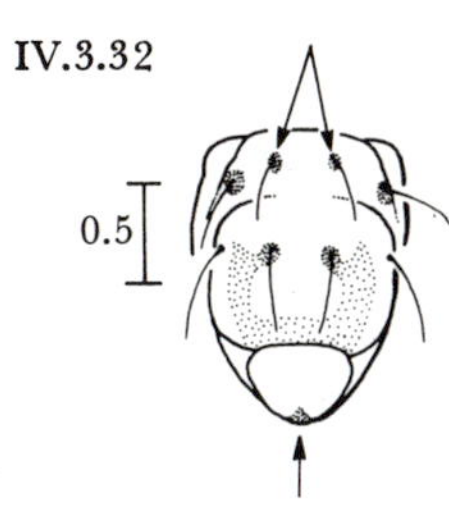

IV.3.33

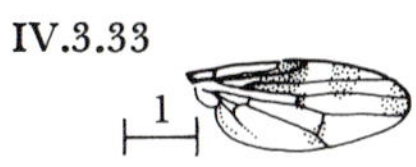

IV.3.34

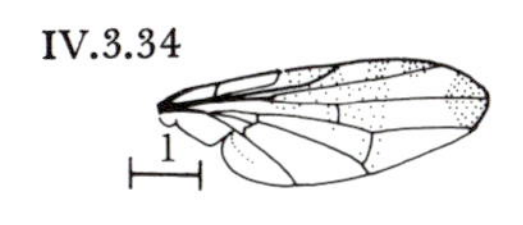

IV.3.35 IV.3.36

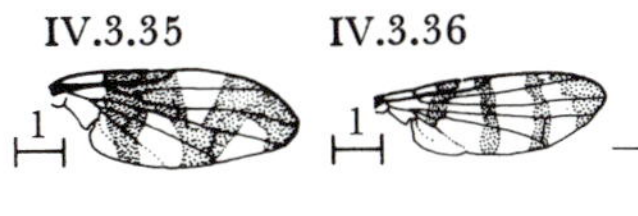

IV.3.37

costal break

Sc

1

IV.3.38

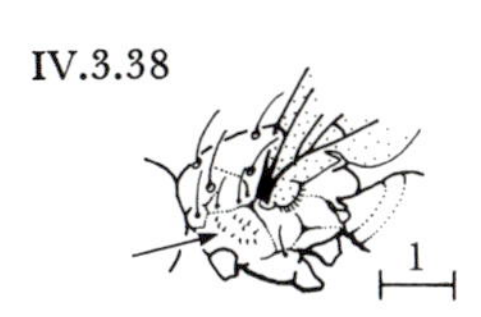

IV.3.39

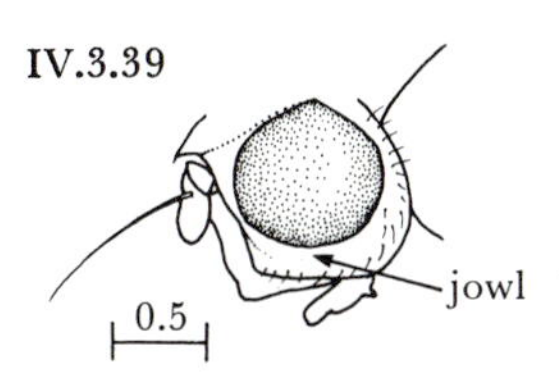

IV.3.40

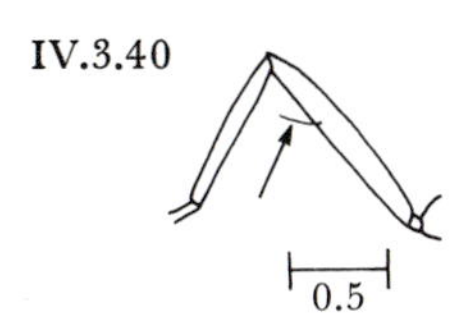

IV.4.1 IV.4.2

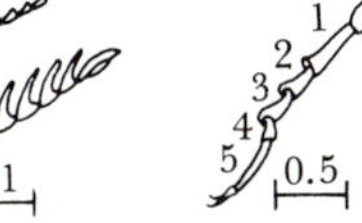

10 Wing markings distinct, extending into posterior ½ of wing (IV.3.33); larva in flower head *Orellia ruficauda* (F.)

– Wing markings faint, usually fading before mid-wing (IV.3.34, though extensions to posterior margin can sometimes be seen using reflected light and dark background); larva in flower head *Orellia winthemi* (Meigen)

11 Wing bands 4, very dark, joined at both anterior and posterior margins (IV.3.35); usually on *Cirsium arvense*; larva in stem gall *Urophora cardui* (L.)

– Wing bands 3 or 4, not joined at posterior margin (IV.3.11) 12

12 3 wing bands, proximal one weak (IV.3.11); common, usually on *Cirsium vulgare*; larva in flower head gall *Urophora stylata* (F.)

– 4 distinct wing bands (IV.3.36); uncommon, on thistles and knapweeds; larva in flower head gall *Urophora solstitialis* (L.)

IV.3.3 Pallopteridae

Check family identification: size >4 mm, with patterned wings conspicuously longer than abdomen; sub-costal vein (Sc) strong without sharp bend, one costal break (IV.3.37); without lower orbital bristles (see IV.3.8).

1 Mesopleura with short hairs (IV.3.38), without a strong black bristle; both cross-veins clouded but no dark cloud on anal vein of wing (IV.3.12); jowls (area below eyes) about as wide as third antennal joint (IV.3.39); hind femur without the bristle as in IV.3.40 *Palloptera parallela* Loew

– Mesopleura bare; dark cloud on anal vein (IV.3.37); hind femur with a strong bristle just beyond middle (IV.3.40); jowls much narrower than third antennal joint *Palloptera umbellatarum* (F.)

(*parallela* is thought to be a synonym of *umbellatarum* by some authorities: see Kloet & Hincks (1975). Other pallopterids may be casual visitors to thistles: to identify see Collin (1951).)

IV.4 COLEOPTERA ADULTS AND LARVAE

To identify superfamilies, see table IV.4.1.

IV.4.1 Cantharoidea

The commonest family on thistles is the CANTHARIDAE (soldier beetles, e.g. *Rhagonycha fulva* (Scopoli): pl. 4.4), often found on flowers where they feed on other visitors.

Table IV.4.1. *To identify adult beetles belonging to five superfamilies of Coleoptera* (If specimen does not fit, try Fowler, 1887–1891; **Joy, 1932**; Crowson, 1956; Chinery, 1976; for larvae, see key IV.5.5.)

	Antennae	Tarsi[a]	Shape, colour etc.	Other characters
Elateroidea: includes click beetles	Toothed or comb-like (IV.4.1)	5-5-5 (IV.4.2)	Body hard; finely hairy above; brown, may have metallic sheen; prothorax with pointed hind angles (IV.4.3)	Hind coxa grooved (IV.4.4); usually with click mechanism (IV.4.5)
Cantharoidea: includes soldier beetles (key IV.4.1)	Thread-like (IV.4.6)	5-5-5 (first segment not shorter than second)	Body soft; elongate; finely hairy above; often red, yellow or black	Tips of elytra squared-off (IV.4.7)
Cucujoidea: includes ladybirds (key IV.4.2)	Thread-like, comb-like or clubbed; club of <4 segments (IV.4.8), not elbowed	5-5-4 or 3-3-3 (IV.4.9)	Body hard; elongate or globular; dull or brightly coloured	
Chrysomeloidea: includes leaf beetles (key IV.4.3)	Thread-like (IV.4.6) or slightly thickened towards tip	Apparently 4-4-4 (IV.4.10)	Body hard; shape various, often brightly coloured or metallic	No rostrum but slight projection on head in Bruchidae
Curculionoidea: includes weevils (key IV.4.4)	Clubbed and usually elbowed (IV.4.11)	Apparently 4-4-4 (IV.4.10)	Body hard; shape and colour various	With rostrum bearing scrobes (grooves) into which scape (first segment) of antenna fits (IV.4.12)

[a]The tarsal formula gives the number of segments in fore-, mid- and hind-tarsi respectively.

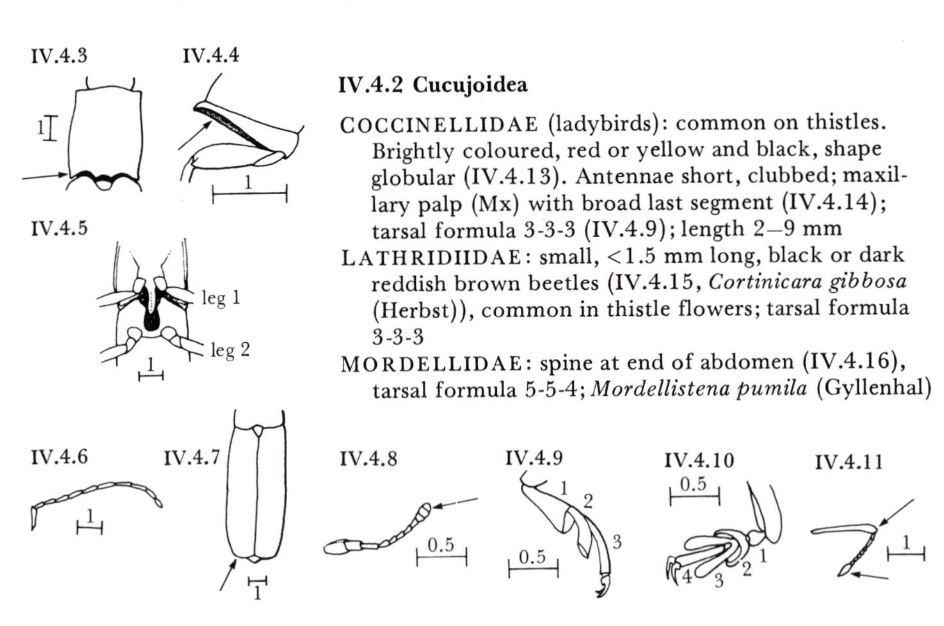

IV.4.2 Cucujoidea

COCCINELLIDAE (ladybirds): common on thistles. Brightly coloured, red or yellow and black, shape globular (IV.4.13). Antennae short, clubbed; maxillary palp (Mx) with broad last segment (IV.4.14); tarsal formula 3-3-3 (IV.4.9); length 2–9 mm

LATHRIDIIDAE: small, <1.5 mm long, black or dark reddish brown beetles (IV.4.15, *Cortinicara gibbosa* (Herbst)), common in thistle flowers; tarsal formula 3-3-3

MORDELLIDAE: spine at end of abdomen (IV.4.16), tarsal formula 5-5-4; *Mordellistena pumila* (Gyllenhal)

IV.4.12

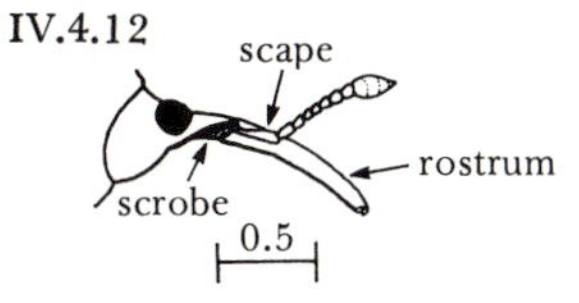

is a stem-borer as a larva; length 3–5 mm; local in S. England

IV.4.3 Chrysomeloidea: includes 3 families

IV.4.13

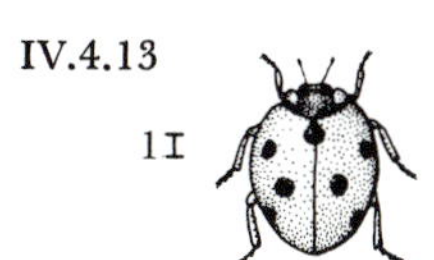

1 Antennae >½ length of body, capable of being reflexed back over body; elytra without regular rows of pits (longhorns) CERAMBYCIDAE

(The larva of *Agapanthia villosoviridescens* (Degeer) (adult, IV.4.17) is a stem-borer; very rare, in S. England only.)

IV.4.14

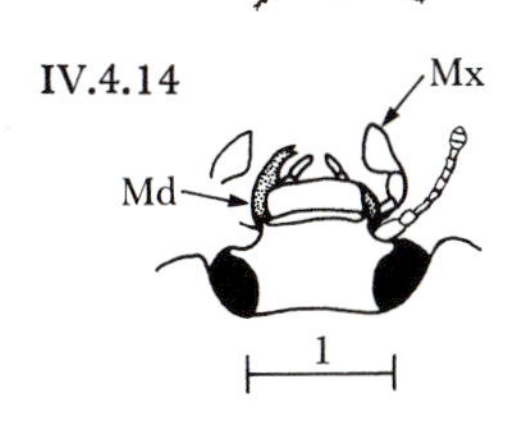

– Antennae <½ length of body, not capable of being reflexed backwards; elytra with or without rows of pits (IV.4.18) 2

2 Antennae thickened towards tip, projection between eyes, elytra often short exposing abdomen (IV.4.19); underside very convex; length 2–5 mm (seed beetles) BRUCHIDAE

IV.4.15

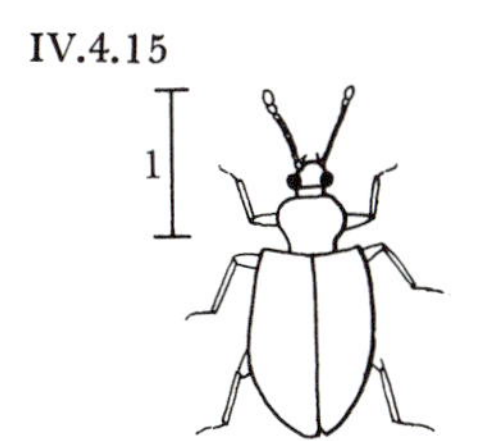

– Antennae threadlike; no projection between eyes; underside flat or slightly convex. CHRYSOMELIDAE (leaf beetles) 3

3 Head completely hidden by prothorax (IV.4.20); underside flat not visible from side; green. Cassidinae (tortoise beetles) 4

– Head not hidden by prothorax; underside visible from side; black or metallic blue 6

IV.4.16 IV.4.17

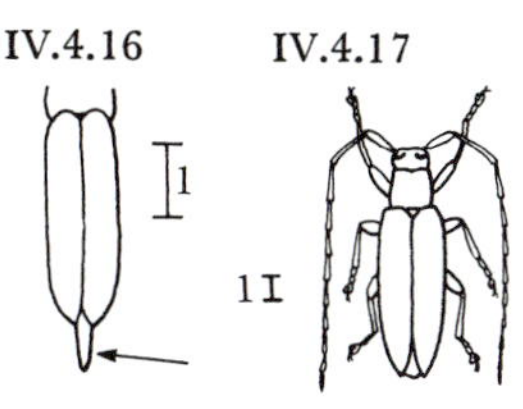

4 Hind corners of prothorax rounded (IV.4.21); length 7–9 mm; occasional visitor to thistles *Cassida viridis* L.

– Hind corners of prothorax angled (IV.4.22); length <8 mm 5

IV.4.18 IV.4.19

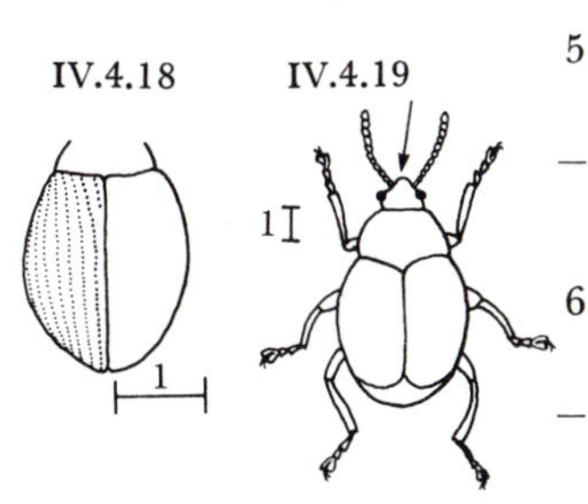

5 Hind corners of prothorax right-angled (IV.4.22); common on thistles *Cassida rubiginosa* Müller

– Hind corners of prothorax obtuse (IV.4.23); usually on knapweeds, local in S. England *Cassida vibex* L.

6 Hind femur much broader than other femora (IV.4.24), used for jumping. Halticinae (flea beetles) 7

– Hind femur not markedly broader than other femora (IV.4.25); cannot jump 12

IV.4.20 IV.4.21

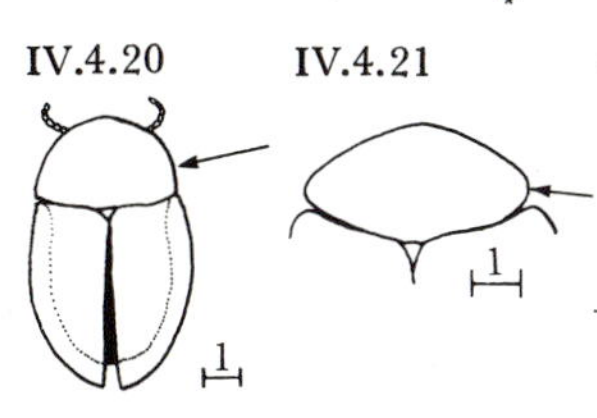

7 Prothorax with short impressed line at each side (IV.4.26, pl. 8.4); beetle less globular, about twice as long as wide (ratio length : width, >1.8); length 3–5 mm *Crepidodera* sp.

– Prothorax without impressed line; beetle globular, less than twice as long as wide (ratio length : width, <1.5) 8

IV.4.22 IV.4.23

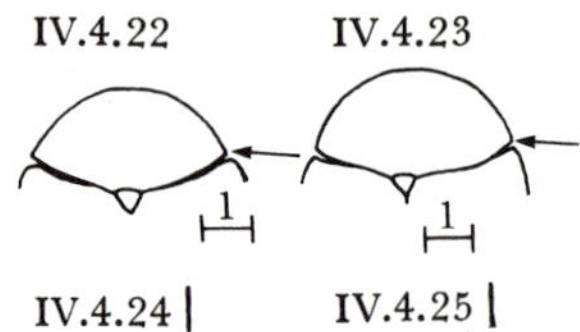

IV.4.24 IV.4.25

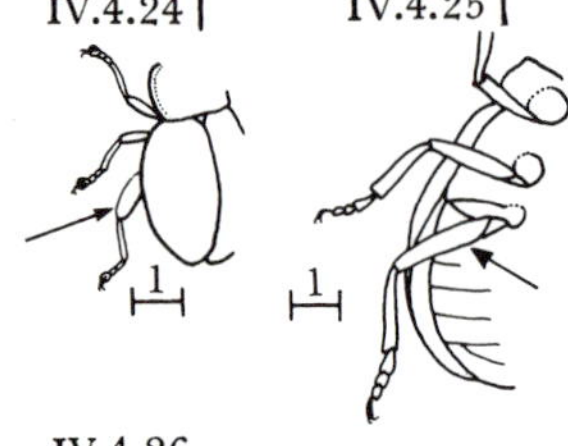

IV.4.26

0.5

IV.4.27

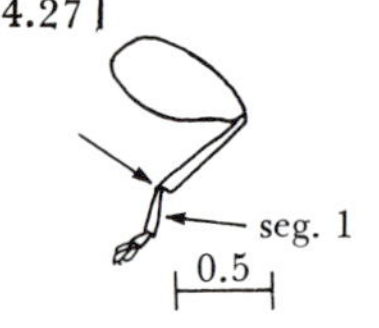

IV.4.28 IV.4.29

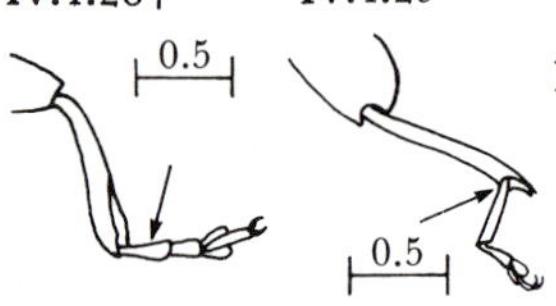

IV.4.30

1

IV.4.31

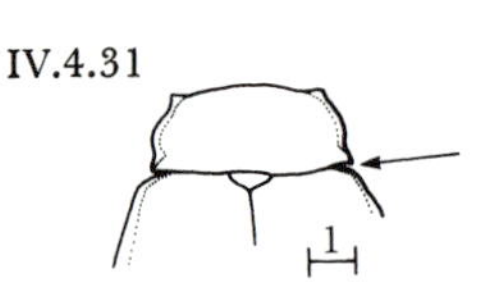

IV.4.32 IV.4.33

1 1

8 Segment 1 of hind tarsus at least twice as long as segment 1 of other tarsi (IV.4.27) — 9

– Segment 1 of hind tarsus <1½ times as long as that of other tarsi (IV.4.28) — 10

9 Length 1.5–2.5 mm, brown; hind tarsus inserted at tip of tibia (IV.4.27); common
Longitarsus luridus (Scopoli)

– Length 2.5–3.5 mm, metallic blue; hind tarsus inserted before tip of tibia (IV.4.29); local
Psylliodes chalcomera (Illiger)

10 Metallic blue or bronze; hind femora black, otherwise legs light brown; elytra with regular rows of pits (IV.4.18, pl. 4.6); length 2.0–2.5 mm
Apteropeda orbiculata (Marsham)

– Light brown; legs yellow-brown; elytra with very small pits not in regular rows (IV.4.30) — 11

11 Length 2.3–3.5 mm; elytra more rounded (IV.4.30); tibiae evenly thickened towards tip
Sphaeroderma rubidum (Graells)

– Length 3.5–4.0 mm; elytra more elongate; tibiae of ♂ abruptly thickened towards tip (pl. 7.8)
Sphaeroderma testaceum (F.)

12 Elytra black, length 6–12 mm; prothorax nearly as wide as elytra at shoulders (IV.4.31); local. Galerucinae
Galeruca tanaceti (L.)

– Elytra metallic blue, length 3–5 mm; prothorax much narrower than elytra at shoulders (IV.4.32). Criocerinae — 13

13 Prothorax indented near middle (IV.4.32); local, on *Cirsium arvense* — *Lema cyanella* (L.)

– Prothorax indented near base (IV.4.33) — *Oulema* sp.
(*O. melanopa* (L.) and *O. lichenis* Voet are found occasionally on thistles.)

IV.4.4 Curculionoidea: Apionidae and Curculionidae (weevils)

Other curculionoid families are unlikely to be found on thistles. The species keyed below feed on *Cirsium arvense* and *C. vulgare*; others may rest on the plants. If a weevil does not fit here, try Fowler (1887–1913) or Joy (1932).

IV.4.34

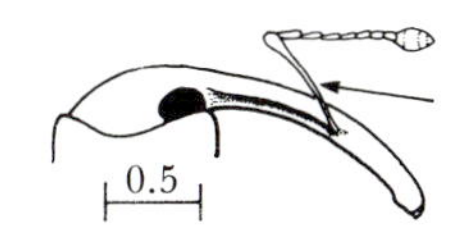

IV.4.35 IV.4.36

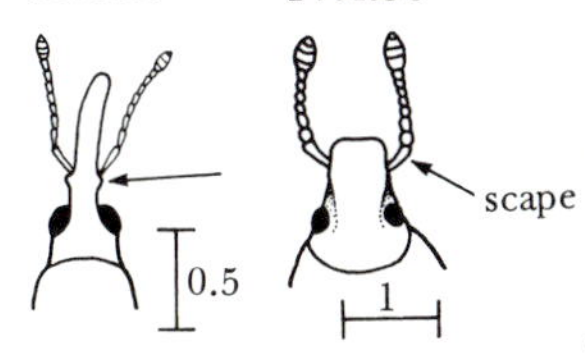

IV.4.37

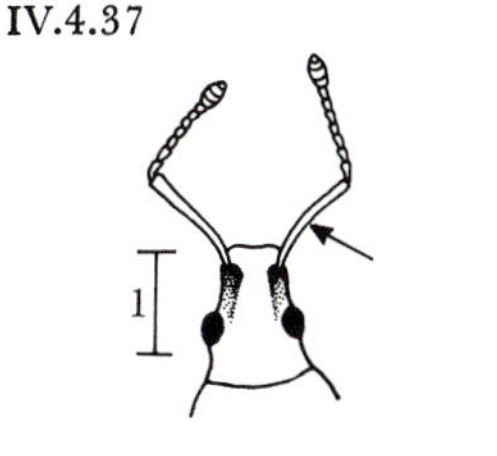

IV.4.38

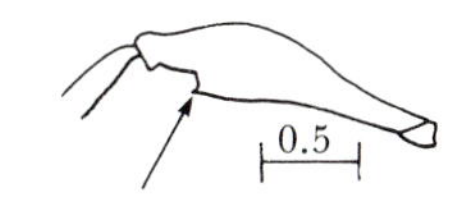

IV.4.39

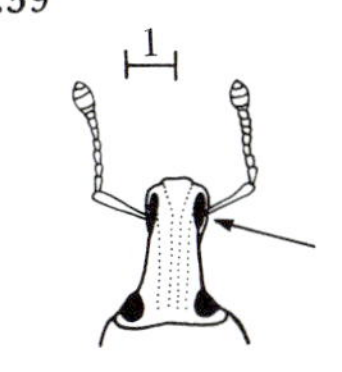

IV.4.40 IV.4.41

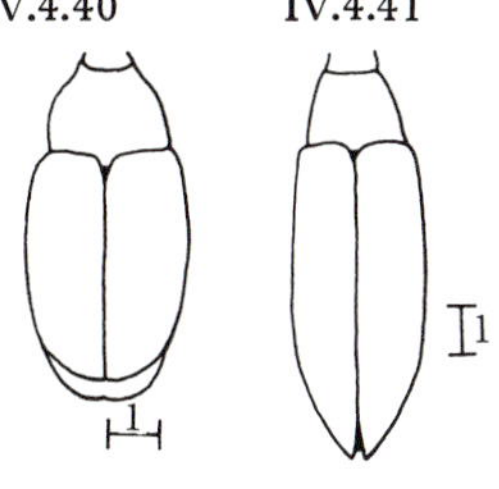

1 Segment 1 of antenna (scape) not or only just longer than segments 2 + 3, antennae inserted in first ⅓ of rostrum (IV.4.12); colour black or blue-black, with hairs but without flattened scales; length 2–3 mm. APIONIDAE 2

– Scape longer than segments 2 + 3 of antenna, antennae inserted beyond mid point of rostrum (IV.4.34); colour various, often with flattened scales; length 3 mm or more. CURCULIONIDAE 3

2 Rostrum with tooth each side at insertion of antennae (IV.4.35); prothorax with smaller pits *Apion carduorum* Kirby

– Rostrum without tooth each side (IV.4.12); prothorax with larger pits *Apion onopordi* Kirby

3 Rostrum broad, <3½ times long as wide (IV.4.36); length >4.5 mm 4

– Rostrum narrow, >4 times long as wide (IV.4.34); length <4.5 mm. Ceuthorhynchinae 9

4 Scape shorter than width of head behind eyes (IV.4.36); femora not toothed. Cleoninae 5

– Scape at least as long as width of head (IV.4.37); femora toothed (IV.4.38) 8

5 Rostrum only just longer than wide (IV.4.36); length 5–6 mm; rare, S. England *Rhinocyllus conicus* (Frölich)

– Rostrum at least twice as long as wide 6

6 Scrobes visible from above (IV.4.39); length *c.* 15 mm; local *Cleonis* (= *Cleonus*) *piger* (Scopoli)

– Scrobes not visible from above 7

7 Rostrum about twice as long as wide; elytra *c.* 1½ times as long as wide (IV.4.40); length 4.5–7.5 mm; rare, S. England and Wales *Larinus planus* (F.)

– Rostrum *c.* 3½ times as long as wide; elytra > twice as long as wide (IV.4.41); length 12–18 mm; very rare, S. England *Lixus algirus* (L.)

8 Prothorax wider than long; length 3–7 mm; common. Otiorhynchinae *Phyllobius* sp.
(The commonest species on thistles are *P. pyri* (L.), *P. roboretanus* Gredler, *P. viridiaeris* (Laicharting) and *P. viridicollis* (F.).)

– Prothorax at least as long as wide; length 7–10 mm; rare. Brachyderinae *Tanymecus palliatus* (F.)

IV.4.42

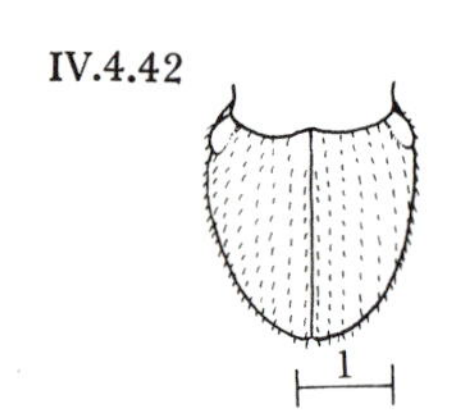

9 Elytra with rows of erect scales (IV.4.42); femora reddish brown; very local
Ceuthorhynchidius horridus (Panzer)

– Elytra without erect scales; femora black with white flattened scales 10

IV.4.43

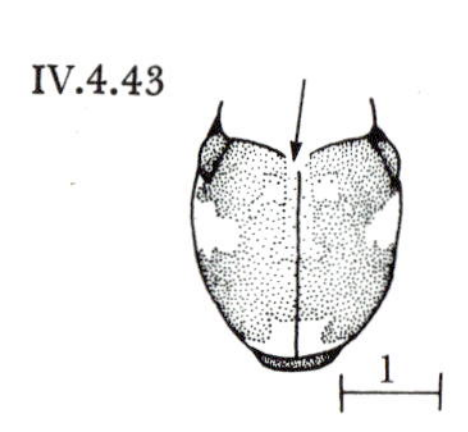

10 Basal third of first interstice (raised line between rows of pits) on elytra with buff or yellow scales, contrasting with adjacent patch of white scales (IV.4.43); bands of white scales on tibiae (IV.4.44, may not be clear on hind tibia); usually on *Carduus nutans*
Ceutorhynchus trimaculatus (F.)

IV.4.44

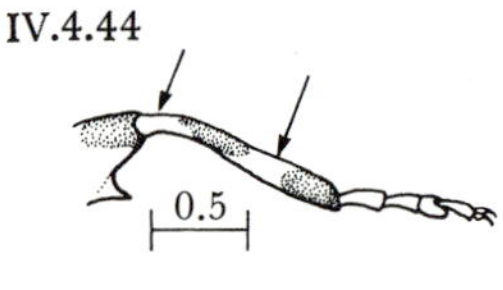

– Elytra including first interstice with white scales only (IV.4.45); white scales on tibiae not in bands; usually on *Cirsium arvense* *Ceutorhynchus litura* (F.)

IV.4.45

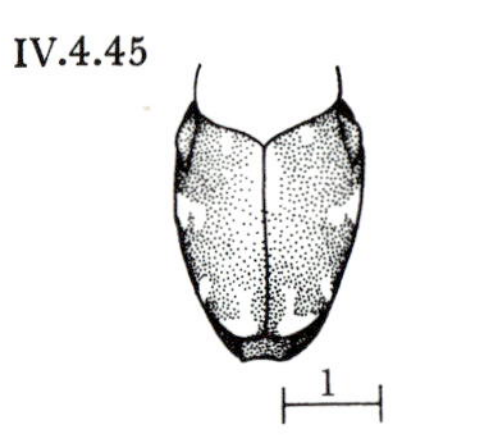

IV.4.5 Beetle larvae

Larvae of two families are common:
COCCINELLIDAE: active, with long legs visible from above (IV.4.46)
CHRYSOMELIDAE: inactive, with short legs usually not visible from above (IV.4.47)

Larvae of *Cassida* spp. (tortoise beetles) are common, broad and flattened, spiny, carrying an umbrella of faeces and cast skins held by spines at posterior end (IV.4.48)

Larvae of *Lema cyanella* (L.) are rare, thick and fleshy, not spiny, carry a faecal umbrella, on *Cirsium arvense* only

Larvae of *Longitarsus luridus* (Scopoli) are long and thin, without a faecal umbrella (IV.4.49)

IV.4.46 IV.4.47

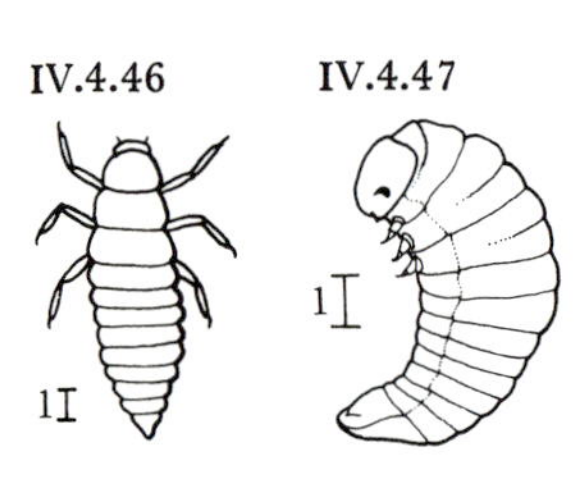

IV.4.48

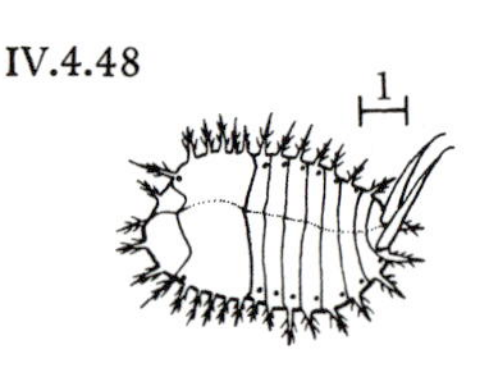

IV.4.49

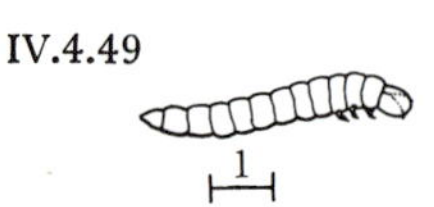

7 Techniques

Apart from a stereoscopic microscope, equipment needed for studying insects on and in thistles is simple. Other essentials are a field notebook and a hand lens, preferably of × 10 magnification. This is easily lost, so tie it to yourself.

Collecting

Many insects feeding on the outside of thistles are inactive enough to be caught directly in a transparent plastic box, a stout cardboard box with a glass lid or a specimen tube. A paint brush can be used to knock the specimen into the container. The advantage of this direct capture is that you know exactly where the insect came from and may be able to watch it feeding, mating, laying eggs etc. before collecting it. A beating tray can be used: spread a white sheet on the ground, bend the thistle over it and rap the stem sharply a couple of times with a stick. The dislodged insects can be sucked up from the sheet with a pooter (fig. 47). For active flying insects, like butterflies, bees and hover-flies, a butterfly net is useful (fig. 48).

For stationary insects on the plant, like aphids, and for those inside thistle heads, leaves and stems, it is best to cut appropriate pieces of thistle with scissors and drop them into polythene bags. If the bags are folded over, stapled and kept in a cool place, plant material and insects will stay fresh for several days. Dead dry heads and stems, including galls and their insects, will keep all winter in a cool frost-free place; if too warm, adult tephritids and parasites will emerge several months early.

Rearing

Most insects in heads, leaves and stems will be larvae or pupae and must be reared to confirm identification. Fully fed tephritids, stem-boring weevils and their parasites can be reared in gelatine capsules (from Eli Lilly & Co. Ltd, Basingstoke, Hants.) which can be pinned with a reference number in boxes lined with cork or expanded polyethylene foam (from Wilford Polyformes, Greaves Way, Stanbridge Road, Leighton Buzzard, Beds.) so that development of individuals can be easily followed. Caterpillars and beetle larvae can be reared in jam jars or tubes containing a layer of sand for pupation. If the larva is still feeding, frass must be removed and the plant material changed regularly to prevent mould developing. The plant will provide enough moisture while green; when pupae are formed, the sand should be kept slightly damp.

For agromyzid larvae in leaf mines, collect a portion of stem bearing the mined leaf and keep it in a pot or tube with a layer of damp sand. Larvae which pupate in the soil will enter the sand; those which pupate in the mine should be removed with a fine paint brush when puparia are formed, and placed on the sand (Spencer, 1976). Wipe away condensation on the sides of the tube, as this can trap emerging flies, and keep the sand damp with a few drops of water every few days. Stem-boring agromyzids need less moisture than leaf-miners, and adults and their parasites will emerge from puparia kept in gelatine capsules. Fully fed larvae of Cecidomyiidae (*c.* 3 mm long) and Pallopteridae will survive over winter in tubes with a 2–3 cm layer of damp sand and closed with nylon stocking fastened with rubber bands.

If you find nymphs of exopterygote insects, like aphids and tingids, it is best to return to the plant a week or so later and collect the adults when they have developed. Other species, like Miridae (and also leaf-feeding caterpillars and beetle grubs), can be confined on the plant in a sleeve cage made of nylon stocking enclosing a leaf or shoot; alternatively, whole plants plus insects can be kept in pots in a greenhouse.

Preserving

Larvae can often be identified alive but it is worth preserving a few so that one becomes familiar with this stage as well as the adult; with practice, some insects may be safely identified from the larva. Soft-bodied larvae and adults, e.g. aphids, should be dropped into

Fig. 47. Pooter.

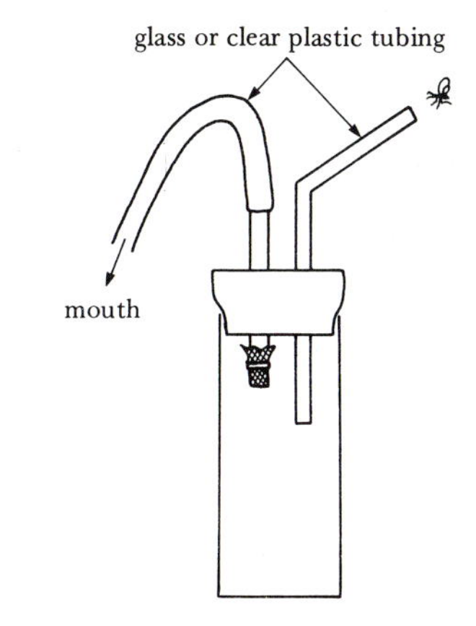

Fig. 48. Butterfly net.

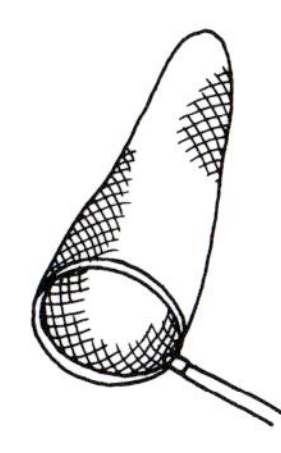

70% alcohol which will kill and preserve them. Alcohol is best also for delicate species, such as Cecidomyiidae and Psocoptera, and for very small ones, such as Collembola and Thysanoptera.

Adults of Lepidoptera, Hymenoptera, Coleoptera and most Diptera should be preserved dry. Ethyl acetate vapour is the best killing agent; wet a piece of tissue with ethyl acetate (chemists will usually supply it) and put it into the tube of insects (use a glass tube as it 'melts' many plastics). After half an hour, most insects will be dead, although the larger tougher bees and beetles may need longer.

Mounting

For ease of identification, adult insects should be mounted on pins (fig. 49; from Watkins & Doncaster, Four Throws, Hawkhurst, Kent) or glued to card points (fig. 50; use a tiny drop of a heavy water-soluble gum) so that as much as possible of the insect is visible (Stubbs & Chandler, 1978). Only Lepidoptera need be set properly before mounting to ensure that the wings can be seen completely (Chinery, 1976). Puparia, pupal cases and cocoons, and mouthparts and genitalia if these have been dissected out, should be glued on to card and pinned with the adult; and galls, leaf mines and damaged stems can be preserved dry or in alcohol or pressed as in a herbarium.

Labelling

All insects at all stages of collecting, rearing, mounting and storage should be adequately labelled. Polythene bags, tubes and boxes containing field collections should also contain labels with date, host plant, locality etc. on them. Larvae must be labelled while being reared, preferably with a reference number listed in a breeding notebook in which a full account of development can be recorded. Reared adults, galls, mines, bored stems etc. should carry the same number and these, and adults collected from the field and pinned, should bear permanent, legible (use indian ink) and complete labels on the same pin as the specimen. The minimum of information on a label should be the host plant or habitat, grid reference, locality and date, together with the name of the insect once it is known (indicate whether this has been confirmed by an expert). Labels for insects stored in alcohol should be written clearly in pencil or indian ink and inserted into the tube; those glued on to the outside often come off.

Insects can be marked with tiny spots of a quick-drying model aeroplane paint such as Humbrol.

How to present your findings

Writing up is an important part of a research project, particularly when the findings are to be communicated to other people. A really thorough, critical investigation that has established new information of general interest may be worth publishing if the animals on which it is based can be identified with certainty. Journals that publish short papers on insect biology include the *Entomologist's Monthly Magazine*, *Entomologist's Gazette*, *Bulletin of the Amateur Entomologists' Society*, and, for material with an educational slant, *Journal of Biological Education*. Those unfamiliar with publishing conventions are advised to examine current numbers of these journals to see what sort of thing they publish, and then to write a paper along similar lines, keeping it as short as is consistent with the presentation of enough information to establish the conclusions. It is then time to consult an appropriate expert who can give advice on whether and in what form the material might be published. It is an unbreakable convention of scientific publication that results are reported with scrupulous honesty. Hence it is essential to keep detailed and accurate records throughout the investigation, and to distinguish in the write-up between certainty and probability, and between deduction and speculation. In many cases it will be necessary to apply appropriate statistical techniques to test the significance of the findings. A book such as Parker (1973) will help, but this is an area where expert advice can contribute much to the planning, as well as the analysis, of the work.

Fig. 49. Insect mounted on a stage (a strip of expanded polyethylene foam or polyporus).

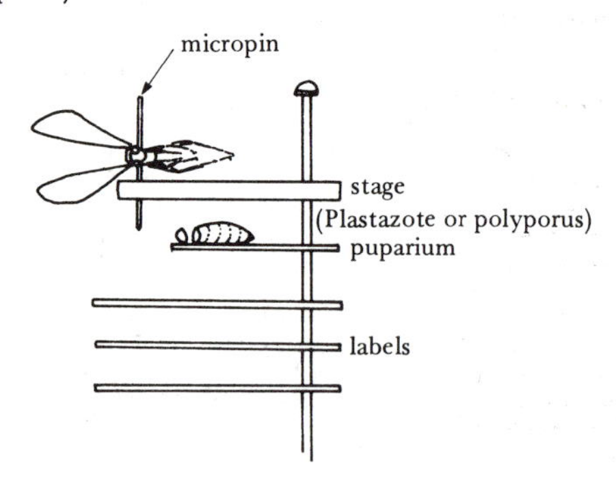

Fig. 50. Insect glued to a card point.

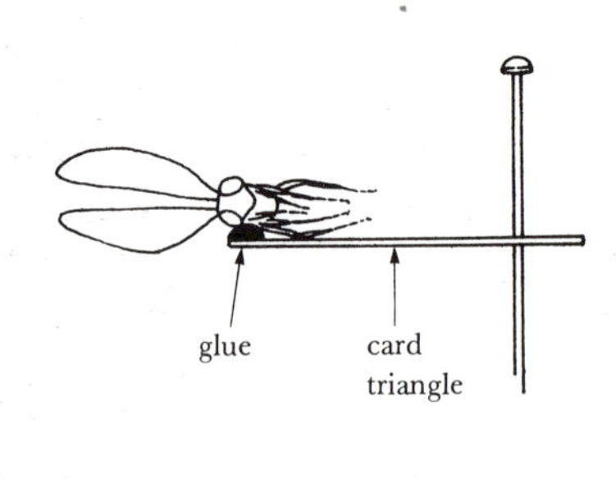

Further reading

Many of the books and journals listed here will be unavailable in local and school libraries. It is possible to arrange to see or borrow such works by seeking permission to visit the library of a local university, or by asking your local public library to borrow the work for you via the British Library, Lending Division. This may take several weeks and it is important to present your librarian with a complete and correct reference. References are acceptable in the form given here: author's name and date of publication, the title and publisher (for a book) or the title of the article, the journal title, volume number and first and last pages (for a journal article).

The *Handbooks for the Identification of British Insects*, published by the Royal Entomological Society of London, may be bought from the Society at 41 Queens Gate, London SW7 5HU, or from E.W. Classey Ltd, PO Box 93, Faringdon, Oxon SN7 7DR. Books marked with an asterisk are out of print.

Askew, R.R. (1968). Hymenoptera: Chalcidoidea. *Handbooks for the Identification of British Insects*, **8**(2b), 1–39.

Askew, R.R. (1980). The diversity of insect communities in leaf-mines and plant galls. *Journal of Animal Ecology*, **49**, 817–29.

Balachowsky, A.S. (1963). *Entomologie appliquée à l'agriculture*, book I, *Coleoptères*, vol. 2. Paris: Masson.

Banks, C.J. & Nixon, H.L. (1958). Effects of the ant *Lasius niger* (L.) on the feeding and excretion of the bean aphid *Aphis fabae* Scop. *Journal of Experimental Biology*, **35**, 703–11.

Barnes, H.F. (1928). British gall midges. II. *Lestodiplosis* Kieffer. *Entomologist's Monthly Magazine*, **64**, 68–75.

*Beirne, B.P. (1954). *British Pyralid and Plume Moths*. London: Warne.

Blackman, R. (1974). *Aphids*. London: Ginn.

Bradley, J.D., Tremewan, W.G. & Smith, A. (1973). *British Tortricoid Moths*, vol. I, *Cochylidae and Tortricidae: Tortricinae*. London: Ray Society.

Bradley, J.D., Tremewan, W.G. & Smith, A. (1979). *British Tortricoid Moths*, vol. II, *Tortricidae: Olethreutinae*. London: Ray Society.

*Buckler, W. (1890–9). *Larvae of British Butterflies and Moths*, vols. I–IX (1899: Vol. IX, *Pyrales etc.*). London: Ray Society.

Buhr, H. (1964). *Bestimmungstabellen der Gallen (Zoo- und Phytocecidien) an Pflanzen Mittel- und Nord-Europas.* 2 vols. Jena: Fischer-Verlag.

Cameron, R.A.D. (1977). Field studies on animal populations: beyond density estimates. *Journal of Biological Education*, **11**, 197–201.

Cameron, R.A.D. & Redfern, M. (1974). A simple study in ecological energetics using a gall-fly and its insect parasites. *Journal of Biological Education*, **8**, 75–82.

Carter, D.J. (1979). *The Observer's Book of Caterpillars*. London: Warne.

Chinery, M. (1976). *A Field Guide to the Insects of Britain and Northern Europe*, 2nd (corrected) edn. London: Collins.

Clapham, A.R., Tutin, T.G. & Warburg, E.F. (1958). *Flora of the British Isles*. London: Cambridge University Press.

Claridge, M.F. (1961). Biological observations on some eurytomid (Hymenoptera: Chalcidoidea) parasites associated with Compositae, and some taxonomic implications. *Proceedings of the Royal Entomological Society of London, Series A*, **36**, 153–8.

Claridge, M.F., Blackman, R.L. & Baker, C.R.B. (1970). *Haltica carduorum* Guerin introduced into Britain as a potential control agent for creeping thistle, *Cirsium arvense* (L.) Scop. *Entomologist*, **103**, 210–12.

Collin, J.E. (1951). The British species of *Palloptera* Fallèn (Diptera). *Entomological Record and Journal of Variation*, **63**, *Suppl.*, 1–6.

Colyer, C.N. & Hammond, C.O. (1968). *Flies of the British Isles*. London: Warne.

Crowson, R.A. (1956). Coleoptera: introduction and keys to families. *Handbooks for the Identification of British Insects*, **4**(1), 1–59.

DeBach, P. (ed.) (1964). *Biological Control of Insect Pests and Weeds*. New York: Reinhold.

Emmet, A.M. (ed.) (1979). *A Field Guide to the Smaller British Lepidoptera*. London: British Entomological and Natural History Society.

Ferrière, C. & Kerrich, G.J. (1958). Hymenoptera: Chalcidoidea. *Handbooks for the Identification of British Insects*. **8**(2a), 1–40.

Fitter, R., Fitter, A. & Blamey, M. (1974). *The Wild Flowers of Britain and Northern Europe*. London: Collins.

*Fowler, W.W. (1887–1913). *The Coleoptera of the British Islands*, vols. I–VI. London.

Graham, M.W.R.deV. (1969). The Pteromalidae of North-West Europe. *Bulletin of the British Museum (Natural History), Suppl. 16*, 1–908.

Griffiths, G.C.D. (1962). Breeding leaf-mining flies and their parasites. *Entomological Record and Journal of Variation*, **74**, 178–85, 203–6.

Harris, K.M. (1966). Gall midge genera of economic importance (Diptera: Cecidomyiidae). I. Introduction and subfamily Cecidomyiinae; supertribe Cecidomyiidi. *Transactions of the Royal Entomological Society of London*, **118**, 313–58.

Harris, K.M. (1973). Aphidophagous Cecidomyiidae (Diptera): taxonomy, biology and assessment of field populations. *Bulletin of Entomological Research*, **63**, 305–25.

Harris, P. (1973). Insects in the population dynamics of plants. In *Insect/Plant Relationships*, ed. H.F. van Emden. *Symposium of the Royal Entomological Society of London*, **6**, 201–9.

Harris, P. (1980*a*). Establishment of *Urophora affinis* Frfld. and *U. quadrifasciata* (Meig.) (Diptera: Tephritidae) in Canada for the biological control of diffuse and spotted knapweed. *Zeitschrift für angewandte Entomologie*, **89**, 504–14.

Harris, P. (1980*b*). Effects of *Urophora affinis* Frfld. and *U. quadrifasciata* (Meig.) (Diptera: Tephritidae) on *Centaurea diffusa* Lam. and *C. maculosa* Lam. (Compositae). *Zeitschrift für angewandte Entomologie*, **90**, 190–201.

Higgins, L.G. & Riley, N.D. (1970). *A Field Guide to the Butterflies of Britain and Europe*. London: Collins.

Jacobs, S.N.A. (1955). On the British Oecophoridae (Lep. Tin.), vol. III. *Proceedings of the South London Entomological and Natural History Society*, (1954–5), 54–76.

Jacobs, S.N.A., Wakely, S., Ford, L.T.,

Brown, S.C.S. & Heath, J. (1978). *Illustrated Papers on British Microlepidoptera.* British Entomological and Natural History Society, occasional publication. (Includes Jacobs, 1955.)

Joy, N.H. (1932). *A Practical Handbook of British Beetles*, vols. I and II. London: Witherby. (1976 reprint, London: Classey.)

Kloet, G.S. & Hincks, W.D. (1964–78). A check list of British insects. Part 1: Small orders and Hemiptera (1964); Part 2: Lepidoptera (1972); Part 5: Diptera and Siphonaptera (1975); Part 3: Coleoptera (1977); Part 4: Hymenoptera (1978). *Handbooks for the Identification of British Insects*, **11**.

*Linssen, E.F. (1959). *Beetles of the British Isles*, vols. I and II. London: Warne.

Meyrick, E. (1928). *A Revised Handbook of British Lepidoptera.* London: Watkins & Doncaster. (1968 reprint, London: Classey.)

Milne, D.L. (1960). The gall midges (Diptera: Cecidomyiidae) of clover flower-heads. *Transactions of the Royal Entomological Society of London*, **112**, 73–108.

Myers, J.H. & Harris, P. (1980). Distribution of *Urophora* galls in flower heads of diffuse and spotted knapweed in British Columbia. *Journal of Applied Ecology*, **17**, 359–67.

Niblett, M. (1939). Notes on the food-plants of the larvae of British Trypetidae. *Entomological Record and Journal of Variation*, **51**, 69–73.

Niblett, M. (1953). Notes on the emergences of Trypetidae. *Entomological Record and Journal of Variation*, **65**, 231–3.

Niblett, M. (1956). Some dipterous inhabitants of thistle. *Entomological Record and Journal of Variation*, **68**, 75–8.

Oldroyd, H. (1954). Diptera: Introduction and key to families. *Handbooks for the Identification of British Insects*, 9(1), 1–49.

Parker, R.E. (1973). *Introductory Statistics for Biology* (Studies in Biology). London: Edward Arnold.

Parmenter, L. (1942). *Palloptera umbellatarum* Mg. (Dipt., Pallopteridae), wing variation and larval habitat. *Entomologist's Monthly Magazine*, **78**, 167–8.

Persson, P.I. (1963). Studies on the biology and larval morphology of some Trypetidae (Dipt.). *Opuscula Entomologica*, **28**, 33–69.

Peschken, D.P. & Beecher, R.W. (1973). *Ceutorhynchus litura* (Coleoptera: Curculionidae): biology and first releases for biological control of the weed Canada thistle (*Cirsium arvense*) in Ontario, Canada. *Canadian Entomologist*, **105**, 1489–94.

Redfern, M. (1968). The natural history of spear thistle-heads. *Field Studies*, 2, 669–717.

Redfern, M. (1972). *Tetrastichus daira* (Walker) (Hym., Eulophidae): a new parasite of *Urophora stylata* (F.) (Dipt., Trypetidae). *Entomologist's Monthly Magazine*, **168**, 51.

Redfern, M. & Cameron, R.A.D. (1978). Population dynamics of the yew gall midge *Taxomyia taxi* (Inchbald) (Diptera: Cecidomyiidae). *Ecological Entomology*, **3**, 251–63.

Richards, O.W. (1977). Hymenoptera: Introduction and key to families. *Handbooks for the Identification of British Insects*, **6**(1), 1–100.

Séguy, E. (1934). *Faune de France*, vol. 28, *Muscidae, Acalypterae et Scatophagidae*. Paris.

South, R. (1961). *Moths of the British Isles*, vols. I and II. London: Warne.

Southwood, T.R.E. (1966) (2nd edition 1978). *Ecological Methods.* London: Methuen.

Southwood, T.R.E. & Scudder, G.C.E. (1956). The bionomics and immature stages of the thistle lace bugs (*Tingis ampliata* H.-S. and *T. cardui* L.; Hem., Tingidae). *Transactions of the Society for British Entomology*, **12**, 93–114.

Spencer, K.A. (1972). Diptera: Agromyzidae. *Handbooks for the Identification of British Insects*, **10**, 1–136.

Spencer, K.A. (1976). The Agromyzidae (Diptera) of Fennoscandia and Denmark, parts I and II. *Fauna Entomologica Scandinavica*, **5**.

Stubbs, A.E. (1969). Observations on *Palloptera scutellata* Mcq. in Berkshire and Surrey and a discussion of the larval habitats of British Pallopteridae (Dipt.). *Entomologist's Monthly Magazine*, **104**, 157–60.

Stubbs, A.E. & Chandler, P. (eds.) (1978). A dipterist's handbook. *Amateur Entomologist*, **15**, 1–255.

Unwin, D. (1981). A key to the families of British Diptera. *Field Studies*, 5, 513–33.

Varley, G.C. (1937). The life history of some trypetid flies, with descriptions of the early stages (Diptera). *Proceedings of the Royal Entomological Society of London, Series A*, **12**, 109–22.

Varley, G.C. (1947). The natural control of population balance in the knapweed gall-fly (*Urophora jaceana*). *Journal of Animal Ecology*, **16**, 139–87.

Varley, G.C., Gradwell, G.R. & Hassell, M.P. (1973). *Insect Population Ecology: An Analytical Approach.* London: Blackwell Scientific.

Wratten, S.D. & Fry, G.L.A. (1980). *Field and Laboratory Exercises in Ecology.* London: Edward Arnold.

Zwölfer, H. (1965). Preliminary list of phytophagous insects attacking wild Cynareae (Compositae) species in Europe. *Technical Bulletins of the Commonwealth Institute of Biological Control*, **6**, 81–154.

Zwölfer, H. (1968). Some aspects of biological weed control in Europe and N. America. In *Proceedings of the 9th British Weed Control Conference*, pp. 1147–56. Delémont, Switzerland: Commonwealth Institute of Biological Control.

Zwölfer, H. (1969). Experimental feeding ranges of species of Chrysomelidae (Col.) associated with Cynareae (Compositae) in Europe. *Technical Bulletins of the Commonwealth Institute of Biological Control*, **12**, 115–30.

Zwölfer, H. (1970). The structure and effect of parasite complexes attacking phytophagous host insects. In *Dynamics of Populations*, ed. P.J. den Boer & G.R. Gradwell, pp. 405–18. Wageningen: Centre for Agricultural Publishing and Documentation.

Zwölfer, H. (1972). Investigations on *Urophora stylata* Fabr., a possible agent for the biological control of *Cirsium vulgare* in Canada. *Weed projects for Canada, Progress Report 29*, 1–20 (unpublished).

Zwölfer, H. (1979). Strategies and counterstrategies in insect population systems competing for space and food in flower heads and plant galls. *Fortschritte der Zoologie*, **25**, 331–53.

Zwölfer, H. & Eichhorn, O. (1966). The host ranges of *Cassida* spp. (Col. Chrysomelidae) attacking Cynareae (Compositae) in Europe. *Zeitschrift für angewandte Entomologie*, **58**, 384–97.

Zwölfer, H. & Harris, P. (1966). *Ceutorhynchus litura* (F.) (Col. Curculionidae), a potential insect for the biological control of thistle, *Cirsium arvense* (L.) Scop., in Canada. *Canadian Journal of Zoology*, **44**, 23–38.

Index